W9-AEP-446

# PROFITS OF SCIENCE

# PROFITS
# OF SCIENCE

## THE AMERICAN MARRIAGE OF
## BUSINESS AND TECHNOLOGY

## Robert Teitelman

BasicBooks
*A Division of* HarperCollins*Publishers*

Copyright © 1994 by Robert Teitelman.
Published by BasicBooks, A Division of HarperCollins Publishers, Inc.

All rights reserved. Printed in the United States of America. No part of this book may be reproduced in any manner whatsoever without written permission except in the case of brief quotations embodied in critical articles and reviews. For information, address BasicBooks, 10 East 53rd Street, New York, NY 10022-5299.

*Designed by Ellen Levine*

Library of Congress Cataloging-in-Publication Data
Teitelman, Robert.
    Profits of science : the American marriage of business and technology / Robert Teitelman.
        p.   cm.
Includes bibliographical references and index.
ISBN 0–465–03983–9
    1. Science—United States. 2. Science—Economic aspects—United States.   3. Science—Political aspects—United States.   4. Technology—United States.   5. Technology—Economic aspects—United States.   6. Technology—Political aspects—United States.   I. Title.
Q127.U6T384   1994
303.48'3'0973—dc20                                                93–33277
                                                                                   CIP

94 95 96 97 ◆/HC 9 8 7 6 5 4 3 2 1

TO CAMILLA

# CONTENTS

# PREFACE

For some time now, I have been reporting and writing about two broad subjects: technology and finance. Like most journalists laboring on a beat, I assembled a very rough-and-ready approximation of the recent history of the two fields. But even when I wrote a book describing their collision—*Gene Dreams: Wall Street, Academia and the Rise of Biotechnology*—I dealt mostly with events that occurred in the very recent past, mostly in the 1980s. Only after completing that book did I begin to wonder whether the particular dynamics among molecular biology, biotechnology, and Wall Street might suggest a more general pattern. How had quantum physics and solid-state electronics interacted in the 1950s? Why did the television-set business evolve so differently from the semiconductor business? What had changed between 1950 and 1990?

In time, I found myself reconsidering three popular explanations that shape the popular discourse on technology. First was the notion that technologies evolve in isolation, unaffected over the long term by their environment. Television developed differently than semiconductors simply because of some inherent technological predisposition, not because of some other element in the environment, such as regulation or finance. Second, culture means everything. Americans are inherently entrepreneurial, which is why we invented venture capital and the transistor, and why we're not so good at building things well (bored, like the brightest child in a dull class). Thus, the network of institutions and arrangements that make up the technological economy reflects some deep cultural requirement—and explains differences with foreign competitors such as the Japanese or Germans. Third is the notion, currently popular among the devotees of the new information age, that capital doesn't matter—or matters less than it used to; or, more accurately, that capital, like water in Los Angeles, will always appear when you turn on the spigot. A corollary is that old-fashioned economies of scale have been banished in the new age.

Each of these totems needs to be, at best, altered, and at worst, dismantled and carted away. Technological innovation and change are complex phenomena that react sensitively to changes in the environment (particularly in finance) and shape the environment. Not only does capital ebb and flow, but different sources of R&D funding—corporate funding, government funding, venture capital, public investors—mold technological progress in different ways. If institutional arrangements reflect culture, why in certain periods of American history have we had corporate concentrations like Japanese *keiretsu,* and why, as recently as thirty years ago, were we enamored of large, hierarchical organizations? In fact, the most striking aspect of the last forty years or so is the remarkable amount of fundamental change that has occurred. Americans have shifted over that time from seeing the economy as the home of the large corporation to viewing it as a vast seedbed for the small. Beneath that shift lies a complex, subtle, and ambiguous relationship between size and innovation. In many ways, this book is a meditation on that relationship.

The key relationship between corporate structure and innovation may appear a dull, technical matter of the dismal science. It is not. The notion of corporate structure really taps much more profound issues of how we govern and organize our lives. The major issue is, at bottom, political. It is one that has, in other manifestations, informed and shaped American political and economic history since Thomas Jefferson, articulating his ideal of a republic of small, pastoral farmers, clashed with Alexander Hamilton's vision of an urban-based, manufacturing nation. Whether we debate antitrust, corporate takeovers, Big Science versus Little Science, the merits of big business and small business, the role of markets, or, indeed, corporate R&D versus the merits of the entrepreneurial firm, we are, in effect, rehashing some of these basic American issues.

For me, this kind of book, so different from conventional journalism, provided particular challenges and occasional deep anxieties. I owe enormous thanks to Steve Fraser, my editor at Basic Books, who showed not only interest in a very sketchy thesis I proposed early on but enormous patience during an unusually long gestation period and—how rare is this these days?—sensible suggestions on how to improve the manuscript that eventually thudded up against his door. My thanks, too, to my colleagues at *Institutional Investor* magazine, particularly my editor, David Cudaback, and colleagues Firth Calhoun and Ken Klee. Not only did working at *Institutional Investor* significantly deepen my understanding of finance, but they were flexible enough to tolerate occasional physical and mental absences. I've bothered a number of friends and colleagues into reading sections of this book: Firth and Ken, Arthur Finch, Dan Cordtz. I am also grateful to Jim Bergland, for his kindness, his conversation, and his office space; to my friend and colleague Herbert Burkholz, for many conversations, not least on the subject of his own book, *The FDA Follies;* and to Susan Burkholz, for excellent research assistance. And finally, to Camilla, for patience and fortitude beyond describing.

*In the beginning, I am saying, was complexity. The essence of life from the beginning was homeostasis based on a complicated web of molecular structures. Life by its very nature is resistant to simplification, whether on the level of single cells, or ecological systems or human societies.*

—Freeman Dyson, *Why Is Life So Complicated?*

# 1

# The Gap between Science and Technology

## Clinton's Progress

In 1946, when historian John Lukacs emigrated to the United States from Hungary, he was struck by the sense that Americans, untouched at home by the great European wars, remained disciples of a gospel of scientific progress that was

> left over from the previous century, consisting of endlessly formulated applications of mechanical utilitarianism, of the corroding cast-iron philosophies of Bentham or Spencer or Buckle or, indeed, Dickens' Gradgrind. . . . The most pervasive American creed was not "The Survival of the Fittest" (that's what Germans believed, not Americans), but of mechanical Evolution, endless Progress.[1]

When writing in the late 1980s, Lukacs found little reason to change that view. In good years and bad years, technology in America had been framed in terms of progress. Was the country moving forward? Or was it sliding backward? Were enough prizes, patents, and new graduates accumulating? Were enough dollars being spent and were they being spent effectively? Had we beaten the Russians to the moon? Were we losing to the Japanese? The roll call of technological triumphs, the accumulation of Nobel Prizes, the sheer size and diversity of the American technological machine (in physicist Robert Oppenheimer's phrase, those "pyramids of invention"), the endless counting of research dollars and Ph.D.'s—all these were the necessary and outward signs of progress and, by extension, America's preeminence in the world.[2]

U.S. president Bill Clinton embraces the gospel of progress referred to by Lukacs, as well as the idea that somehow America has gone wrong. During the 1992 campaign, Clinton argued that America was going through a wrenching transition to a new age dominated by information technologies and by a class of white-collar workers; his was a platform built upon ideas promoted by his old Rhodes scholar friend, Harvard professor and current secretary of labor Robert Reich. From that fairly common worldview the Clinton campaign built its case for improving education and retraining (to shift blue-collar workers to new, information-oriented jobs), managing trade (to protect high-technology industry), and building a twenty-first-century technological infrastructure. The campaign also built a case for promoting R&D and small-business tax credits, for dampening the predations of the financial market, and for banishing "short-term" thinking and discouraging the flight of technology and jobs overseas. Although by 1992 Reich had seemingly abandoned industrial policy, for which he was once a leading champion, the Clinton campaign talked of "protecting" certain key industries, particularly the semiconductor and computer establishment, while also nurturing technology in small companies.

By April 1992, when Clinton announced his budget, the practical thrust of his administration became clearer. The administration slashed the budgets of nine of the nineteen research institutes at the National Institutes of Health (NIH), reducing areas of basic science in order to fund "target" areas of applied science: AIDS, minority health, and breast cancer. On the other hand, both the National Aeronautics and Space Administration (NASA) and the National Science Foundation (NSF) received relatively large increases, mostly for areas that the administration believed would bring applied, and commercial, results, from high-speed planes and high-speed computers to programs supporting science and engineering education.[3]

Despite the shiny new rhetoric, few of the ideas rationalizing the Clinton plan are very new. In austere times, politicians, like corporate executives, often sacrifice basic research to save applied science; after all, researchers at the NIH will not fund campaigns—companies will. More important, the underlying debate over the Clinton program—the notion that a political elite should, or could, target certain sciences and technologies for support—goes back at least to 1945, when Congress and the Truman administration began to try to develop policies for postwar technology management. The debate between Vannevar Bush, the wartime head of the Office of Scientific Research and Development (OSRD), and West Virginia senator Harley Kilgore touched on most of the major themes that have been played over and over ever since. Bush argued that scientists themselves should control the scientific agenda and that funding should be limited only to the hard, basic sciences; he did not think that companies needed help in developing "technology." Kilgore was more optimistic; he argued that the government should support social sciences as well, that it should target its money toward resolving social and economic difficulties. Kilgore believed that science had matured sufficiently during the war that it could be channeled and controlled by the body politic at large.

Clinton hews to the Kilgore line, though now economic and trade wars predominate over social and developmental issues. One of the more striking aspects of the current debate is how little reference to historical experience, particularly to events since World War II, is ever made. Although there is much talk about changing the system, there is little discussion about why the technological economy evolved as it did. In this book, I attempt to fill that gap by asking a number of more general questions. What have we learned since World War II about the management of technology, about the relationship between science and technology? What has been the effect of government regulation and government targeting of resources? How has the technological economy changed, and what effect has that change had on business and the larger economy? What are the constituent elements of the technological economy, how do they interact, and how do they shape the commercial structures that transmit high science and applied technologies to the quotidian world?

## The Postwar Technological Economy

The popular heroes of the postwar technological economy have naturally been scientists and engineers. But while the masters of quantum mechanics, molecular biology, and innumerable other scientific disciplines have played an instrumental role, they are hardly the only characters in the story. American technology today is as much the creation of bankers, venture capitalists, pension fund managers, and investors (not to mention dreaded politicians, policymakers, and bureaucrats) as it is of scientists and engineers. In fact, it is the active and extensive interaction among such different worlds (the lab, the line, the university, Wall Street, Washington) that makes this age unique. Financial and policymaking figures provided the capital that fueled the machine and, in large measure, determined where it would travel. The amount of money they spent was unprecedented, if often exaggerated. Their success was obvious, if often distorted. For at least three decades after World War II, the technological economy seemed to operate in happy perpetuity. Science and technology produced invention and innovation, which generated economic growth, which in turn spawned the surplus to feed more technology. Only when productivity growth stalled in the mid-1970s did doubts take hold.

The story of the interaction between science and money is complex. During World War II, the federal government, abandoning its prewar disregard for science, opened its pocketbook and nudged the large corporation aside as the central source for R&D funding, particularly for basic science (corporations continued to spend for research, of course, although their relative importance waned). Through the Defense Department and the NIH in the 1950s and 1960s, the government built up American academic science and bankrolled a network of national laboratories. Professional venture capital surfaced soon after the war. By the 1970s, institutional investors, guardians of deep pools of private savings, were pumping money into venture capital funds at the encouragement of Washington. By the 1980s, with federal aid at a plateau, investors in the public markets—from equity to high-yield debt—had grown convinced that the rewards of technological start-ups exceeded the obvious risks, and they leapt willy-nilly into the game.

This steady growth and decentralization of funding fed a sharpening spirit of competition between two classes of corporations: the established firm and the technological start-up. Small companies have traditionally struggled, usually ineffectually, against larger competitors. In retrospect, the 1980s tendency to cast all American business history as one long triumph of the entrepreneur smacks more of wishful thinking than of fact. Historically, from the 1890s to the 1950s, the corporate division of the technological economy was dominated by a handful of giant companies: AT&T, General Electric, RCA, Westinghouse, Du Pont. This dominance was anchored by economies of scale: first, particularly in the earlier decades, a laissez-faire capitalism that favored the competitor with the conservative virtues of deep financial resources and long horizons, and second, an industrial economics that rewarded large, centrally managed enterprises exploiting mass-production techniques. Except for a brief interlude in the 1920s, capital was limited, affording those with access to it (the cheaper the better, and nothing is cheaper than internally generated capital) enormous reach and advantage. Large corporations did most of the spending on science and technology, deployed their financial muscle to build patent fortresses, heavily influenced the government's R&D and regulatory agendas, and, again through the power of the purse strings, held the greatest sway over academia. By the late 1950s, the balance between the two classes of corporations was shifting. The federal government had broken the corporate hegemony over R&D. Academic science swelled. Federal laboratories and funding bureaucracies such as the NIH, the Defense Advanced Research Program, the Department of Energy, and NASA dominated their respective fields. And smaller companies fed off the surfeit of publicly financed research. Moreover, some of the new technologies, particularly in electronics, did not seem to share the economies of scale of the automobile or steel industries.

In this new environment, the once persuasive case for vigorous antitrust enforcement faded; in certain industries, the once central role of the patent declined. New companies—Texas Instruments, Digital Equipment, Intel, Microsoft, Genentech—surfaced in great waves, many with a technological mission and a zeal for competition. The ready availability of capital bought these new players more time to commercialize new technologies, to develop new products and skills (marketing, manufacturing, management) that had traditionally allowed large corporations to preserve their positions in a dangerous, unstable world.

## Science, Technology, and Capital

This book takes a journey through the past forty-odd years of the American technological economy. Throughout these pages, I have attempted to weave perspectives from two vantage points: the micro world of individual technologies (and companies) and the macro environment of legislation and regulation, economic thought and reality, and financial innovation and change. One casts light upon the other; both perspectives are necessary even to begin to gain a rounded view of this era.

The central characters in this story are a handful of prominent technologies that came of age in the decades after World War II. The postwar technological economy was born during the war at a pair of government-financed projects involving civilian scientists, mostly physicists. The Manhattan Project, which built the first atomic bombs, and the Radiation Laboratory (Rad Lab), which developed sophisticated radar systems, both demonstrated how science, notably, quantum physics, could be effectively married to an ongoing, accelerated product development program. Radar in particular had a major influence, technically and organizationally, on the postwar style of R&D. In the early 1950s came the magic of television, the first great postwar consumer electronics product. But as a willing hostage to Federal Communications Commission (FCC) standards, television evolved haltingly and steadily shrank into oligopoly until the Japanese so rudely arrived. The transistor, on the other hand, seemed to rewrite the old industrial rules and attracted waves of new companies that buried not only the vacuum tube but the oligopoly of vacuum-tube makers. If television was static, like steel or cars, transistors were fluid, and this fluidity was a trait passed on to its successors: integrated circuits and microprocessors and its close relations, computers and telecommunications systems.

The story of postwar pharmaceuticals is also one of oligopoly formed in a static environment only to be confronted suddenly with accelerating technological change—in this case, from the biotechnology industry in the 1980s. At that time, investors eagerly poured money into biotechnology, believing that returns would resemble the semiconductor and computer stocks of the 1960s and 1970s. Alas, the biotechnology experience was fundamentally different from that of transistors and integrated circuits, which investors soon discovered to their chagrin.

All of these technologies possess their own individual dynamics—in their life cycles, their capital needs, the time they require to mature. In fact, as technologies evolve, they make varying demands, particularly in terms of capital, call for novel organizational strategies, and possess different potentials for profit. A semiconductor firm, with rapid product turnover, will require an organizational structure far different from that of a drug company, which may introduce a major product every three to five years.

It is a mistake to view technologies as if they were isolated phenomena, rising as if by spontaneous combustion. Rather, they are built on a scientific base, shaped by the financial environment. Science, like technology, operates in cycles, though these cycles tend to unfold over so long a period that they blend into the background. The sciences are also individual: molecular biology did not unfold in quite the same way, at quite the same rate, as did quantum mechanics. Generally speaking, the ease with which a technology is commercialized varies with the maturity of its underlying science. We can judge that maturity only through circumstantial evidence. Has this particular branch of science spun off commercial products already? How much of a consensus exists on fundamental theoretical issues? Where does the science lie on a spectrum between Baconian empiricism (try this, try that, see if it works) and effective model building (extrapolating behavior from an already established theoretical model)? At the start of a project, can one be fairly certain when it will be completed and what it will look like? Can the engineers master it? Have the lines between engineering and science begun to blur?

At times, the nature of the underlying science directly shapes organizational structures and financing patterns. The bureaucratic structure of most major drug companies and of drug regulation was formed, in part, as a response to the stubborn uncertainties of drug discovery and biology. The biotechnology firms thought they had the means to bridge those unknowns—after all, they called the discipline genetic *engineering*—and organized themselves much as chip or computer companies were organized: with scientific leadership, lean professional management, and funding from Wall Street.

The availability, conditions, and price of capital combine to play a major, often discounted, role in determining the arc of individual technologies. This postwar period, for instance, has been characterized by an increasing intensity of capital for technological companies, which had a significant effect over time. The scarcity of venture capital in the early 1950s meant that television makers came predominantly from the ranks of prewar radio companies; the flood of equity capital in the 1980s floated many new computer and semiconductor firms, despite daunting costs. Research-based companies that cannot generate their own cash find themselves hostage to the whims and wiles of the capital markets, as biotechnology discovered. Each financial source, each financial instrument, differs in its demands; obviously, the structure of defense contractors, funded by the government, differs radically from an industry funded by Wall Street.

The fact is that some companies and technologies appear on the stage to raise capital in times when the markets are flush, generous, and stable, and they create one kind of industry. Others must struggle through more austere or volatile times, creating a different kind of industry. Biotechnology could never have been born in the 1950s and would have been a far more modest phenomenon if it had appeared in the 1960s.

## Stages of Growth

The interplay among science, technology, and capital produces a strange music, like wind chimes on a blustery day. Is it just noise, or is there a harmony, a pattern to the song? Economists have been struggling to define the relationship between innovation and organizational structures since the late Harvard economist Joseph Schumpeter raised this issue during World War II. What kinds of companies—large or small, atomistic or monopolistic—drive innovation most effectively?[4]

I believe industrial structures are less cause than effect, that the visible structure of corporations—large or small, hierarchical or decentralized—determines innovation less than the surrounding technological and financial environment. That environment, in turn, hinges upon the changing relationship among science, technology, and capital. Here is a model of how a steadily narrowing gap between science and technology actually alters industrial dynamics and structures: Begin with a condition in which a science and its related applications are widely separated—say, between biology and drug discovery in the 1950s. In these circumstances, even fundamental scientific advances—for instance, the elucidation of DNA in 1953—fail to generate products and have little effect on the engineer or scientist operating in corporate R&D. The gap is simply too wide to traverse effectively. In such a situation, technological innovation is slow and halting, and it requires large amounts of capital to make incremental improvements.

Such conditions produce secondary phenomena. Because the scientific map is a mass of unknowns, rational extrapolation (particularly the use of mathematics) is difficult to accomplish; empiricism rules. Engineering practices dominate corporate research, but it is a form of engineering that depends heavily on accepted wisdom, on rule books and rules of thumb. The industry tends to be dominated by a small number of large corporations that have the financial resources to operate. Patents and other means of protecting, or hoarding, intellectual property become very important. Market control may come from dominating innovation, through patents, or from retaining a stranglehold on marketing or distribution channels. Antitrust issues are thus inflammatory.

Once the gap between science and technology begins to narrow, various forces are unleashed. Early on in this process, the large corporation may well increase its hegemony. After all, only the largest corporation will have the resources to exploit the small number of opportunities tossed up by scientific advance, as Du Pont exploited nylon or as IBM dominated the mainframe computer business. At this stage, the cost of research and development is high partly because corporations must pursue extensive scientific research and engineering at the same time, as drug companies today pursue molecular biology and traditional chemical screening simultaneously.

As the gap continues to narrow, however, the advantages of capital erode. Science begins to build more powerful models, allowing engineers and applied scientists to extrapolate from a given set of conditions. Scientists and engineers begin to think alike; the lines between applied and pure science blur, as they did for the scientists at the Rad Lab, at the Manhattan Project, and at Bell Labs with the invention of the transistor. The pace of technological change accelerates. Costs fall not only because of capital-saving economies but because the time required to move from lab to market dramatically shrinks and the risk of slamming into a previously unknown pitfall declines. Suddenly, the particular strengths of the smaller company—speed, flexibility, sheer entrepreneurial drive—surface. And with capital advantages shrinking, with the science increasingly accessible, the game becomes one of research productivity, a game that the smaller company may even occasionally win. Thus, industries can undergo wracking changes, even (though rarely) true revolutions. With competition rising, antitrust loses its raison d'être. With change accelerating, owning intellectual property diminishes in importance. Around these industries arises the gospel of the entrepreneur in all its many manifestations. And, increasingly, traditional barriers separating science (usually in academia) and commerce break down, often accompanied by fractious controversy.

In time, the gap closes even further. With scientific knowledge widely diffused, there are few advantages to being particularly creative technologically. Commoditization sets in. Now competition occurs over control of marketing, distribution, intellectual property. Capital costs again increase, as innovation grows more expensive and the extent of marketing (the need to differentiate between nearly identical products) mounts. That deadly word, *maturity,* is whispered. This stage, of course, closely resembles the first stage, in which the gap seems to be very wide. In fact, the two stages are identical, because the only way to set the cycle moving again is for the company to tap a scientific area that is again closing the gap between its technological counterparts—much as Japanese television makers in the 1960s shifted from vacuum tubes to transistors, sealing the U.S. industry's eventual demise.

There is one major variable to this general scheme, one of enormous significance to the story of the postwar technological economy. The key factor that can alter these relationships is the presence or absence of capital. Cheap, available capital can shrink the gap between science and technology, accelerate change, undermine the hegemony of the large corporations, and occasionally hold off the shadow of maturity. Large reserves of inexpensive financing can allow smaller, newer companies to exploit scientific possibilities earlier. Thus, one major focus throughout this book is on describing the changing role of financing and the markets; I believe that the presence of proliferating sources of capital distinguishes the postwar era from earlier periods.

That plenitude of capital also helps create the appearance, if not the reality, of technological revolutions. In the postwar years, nearly every technology, major as well as minor, attempted to don the mantle of the revolutionary. Fund-raising, no matter the source, involves the artful use of public relations, with its concomitant disdain for precise definition and weakness for hyperbole. In a cynical age, to label a product or technology "evolutionary" (which, of course, most are) is to dismiss it with faint praise. To claim "revolutionary" potential loudly enough attracts attention, a prerequisite for any fund-raiser or financier. The true revolutionary technological breakthroughs have been rare. Arguably, there has been but one, albeit one with enormous implications: the semiconductor.

Is there some way to rescue the concept of "technological revolution" from hype? If pure science can be divorced from use, technology cannot. American technology exists to be exploited commercially. It has to be funded, manufactured, and marketed. It is, as much as real estate, a form of property. And, like real estate, it is an asset that gains or loses value. The speed and arc of the development of technology can thus best be traced, like the presence of a cook in the kitchen, through the crash and bang of the commercial world. Herein lies a crucial difference between scientific and technological revolution.

A scientific revolution occurs when a fundamental notion is overturned. On the other hand, the most obvious sign of a technological revolution appears only when the corporate users of a mature technology—an ancien régime of vacuum-tube makers or horse-drawn-buggy manufacturers— are overthrown by firms wielding new and powerful techniques. In the scheme of entrepreneurs and corporations, a technological revolution occurs when the entrepreneurs win.

There is, of course, a slippery element to this definition of *revolution,* but it offers an escape from the even slipperier popular version of technological revolution. It fixes the phenomenon firmly on economic criteria (as opposed to social criteria) and places it in a real-world context: the world of companies, large and small, and of financiers, managers, and markets.

## Christians and Pagans

At bottom, all commentators on technology fit into two camps. The first, the Christians, believe that history has become linear, that progress moves like an arrow into the future, that we have made a sharp, revolutionary break with the past. This party, which includes most of the futurist community and most technological optimists, makes a number of compelling arguments. At the heart of their scheme is the notion that a certain self-consciousness toward scientific research, technological development, and business management has freed people from the harness of the past, from the historical tendencies of birth, growth, and decay. The machine of progress has learned to drive itself.

Then there are the pagans, who believe in the inherent cyclicality of history and, more specifically, of technology and the economy. Instead of revolution, the pagans see evolution; the past merges with the future. Science and technology, not to mention the economy at large, remain cyclical. Despite a clear acceleration over the past forty years—and a long ascendance of living conditions over four centuries—economic progress as we know it has less to do with mastering technology's innermost secrets than with the effect of an immensely larger scale and more capital on the technological enterprise.

This distinction between technological Christians and pagans is especially relevant in the 1990s—an age that has been widely declared as a transition to a higher, grander, more productive era. Are we now in a new world, are we new men and women able to generate self-sustaining technological progress? Or will microelectronics and all its spinoffs, and biotechnology and its kin, eventually age, mature, decelerate, and again produce an economy of large, centralized corporations? The apostles of a new age—Peter Drucker, George Gilder, Alvin Toffler—extrapolate from the present, drawing the upward-curving line into the future. Their beliefs elude testing because such beliefs lie in the future; that, of course, is part of their appeal. To be a Christian is to engage in a faith; to be a pagan is to feel the frictions and imperfections of history dragging every soaring satellite back to a welcoming, if imperfect, blue-green earth.

# 2

## The World of Tomorrow

### The Triumph of the Large Corporation

The 1930s was a decade of frightening economic distress, mounting international tensions—and technological wonders. Du Pont wove the first synthetic fibers. RCA developed the first fully electronic television. The DC-3, the world's first large airliner, was flown. Plastics began to appear in products. Physicists probed the nucleus of the atom with whirling cyclotrons. The first antibiotics were developed. By the 1930s, cars had begun to shed their Model T boxiness for a streamlined look, and radio, with its tangle of wires and glowing vacuum tubes, dominated home entertainment. Against the gloomy backdrop of joblessness and poverty, technology shone brilliantly, particularly at a series of fairs spaced like lamps throughout the decade: Chicago's 1933–34 Century of Progress exhibit, San Diego's 1935 California-Pacific exposition, San Francisco's 1939 Golden Gate show. These celebrations climaxed in 1939 with the grandest effort of all, New York City's World of Tomorrow exhibit, a fair that cost $150 million to build and sprawled across 1,126 acres of Flushing Meadows, Queens, a former dump site.

The World of Tomorrow was larger, grander, and more ambitious than earlier fairs of the 1930s. It embodied a spirit of technological enthusiasm that merged with liberal New Deal sentiments. Its full title was "A Happier Way of American Living through the Interdependence of Man, and the Building of a Better World of Tomorrow with the Tools of Today," windy sentiments that novelist and historian H. G. Wells echoed in an article he wrote for the *New York Times:* "The World's Fair in New York is to differ from most World's Fairs in being a forward-looking display. Its keynotes are not history and glory but practical anticipation and hope. . . . It is arranged not indeed as the visible rendering of a utopian dream—there is to be nothing dreamlike about it—but to assemble before us what can be done with human life today and what we shall almost certainly be able to do with it, if we think fit, in the near future."[1]

Despite Wells's claims, the fair had a dreamlike quality, from its central symbols, the Perisphere and Trylon, to the pavilions along flower-bedecked avenues. The fair particularly showcased aggressive city planners, such as New York City commissioner of parks Robert Moses, and industrial designers, such as Raymond Loewy and Norman Bel Geddes. Fair organizers decreed that pavilions represent no known style and that they "transcend" architectural history and tradition.[2] The future would be qualitatively different from the past. "Picture yourself in a building unlike any you have ever seen—a great gleaming sphere entirely covered with stainless steel," announced an advertisement for the U.S. Steel exhibit.[3] The Futurama exhibit of General Motors, with its sophisticated sound system and striking dioramas of whizzing highways, became a sort of representative vision. This pavilion, designed by Bel Geddes, painted a future, circa 1960, that was an amalgam of elevated highways, sweeping geometric shapes, streamlined cars. No dirt, no grime, few people.

Despite the dreamlike boldness of its surface, the fair reflected commercial realities. At center stage stood those giant U.S. corporations that dominated American economic life: mighty General Motors, Chrysler, and Ford; ubiquitous American Telephone & Telegraph; along with Radio Corporation of America, General Electric, Firestone, Goodrich, Borden, Continental Baking, U.S. Steel, Du Pont. These were the largest corporations in the American economy. For the most part, the products manufactured by these companies required complex production processes and were sold through large, national distribution and marketing networks. The technologies that fueled these firms operated on massive industrial scales, required large and continuing capital expenditures, and produced large and stable streams of profits. Generally, the most advanced of their core technologies—electronics (radio, television, telephones), electrical machinery, the automobile, steel, rubber, chemicals—had been developed in the late nineteenth century, the so-called Second Industrial Revolution, and, if not necessarily mature, had been on the scene for some time.

These corporations had an air of permanence about them. Between 1921 and 1946, the rate of turnover among America's top two hundred industrial corporations sharply declined, despite the rigors of the depression.[4] A major factor in that stability was the lock these companies had on new technology, not only through manipulation of the patent system by which they co-opted individual inventors but through a mastery of industrial research. The development of systematized industrial research began in the electrical and chemistry industries in the late nineteenth century (notably at General Electric, AT&T, and Du Pont) but had subsequently diffused throughout the corporate sector. Small businesses, all but invisible at the New York fair, lacked that accessibility to the latest technological development. They were, as a group, in a sorry state, battered by the seemingly endless slump, starved by a dearth of capital, engaged in a constant struggle to survive. By 1941, there were some three million companies in America, 90 percent of them defined as small enterprises. But, as one contemporary study reported, "the trend since the beginning of the century is for small businesses to become smaller and smaller and for big business to become bigger and bigger." During the depression, "small businesses were displaced from entire industries and elsewhere experienced competitive deterioration."[5] Small businesses continued to survive, even prosper, in industries such as retail and services, but they had all but disappeared from most technologically based industrial sectors.

The large corporation also dominated other sectors of the technological economy. Both academia and the federal research establishment slowed the growth of funds for basic research during the depression. Thus, R&D was increasingly confined in the public mind to a sort of science divorced from real life or to some dull, if remarkably productive, agricultural research. In 1940, the federal government spent $74 million on R&D, $29 million of it going to the Department of Agriculture, about $3 million more than was slated for the Department of War. Throughout the 1930s, the federal government provided 12–20 percent of R&D funds; private industry generated fully two-thirds of the total.[6] During this period, the gap between applied research and pure science widened. Expenditures for (mostly applied) research by the government and corporations more than doubled during the depression; outlays by universities and colleges (mostly for basic), on the other hand, grew by only half. Also, spending by the major supporters of basic science—the elite endowed research institutions like the Rockefeller Foundation and the Carnegie Institution—fell. Although applied research received six times as much funding as basic science in 1930, it received ten times as much in 1940.[7]

The domination of industrial R&D was reflected in contemporary attitudes. Technology was the stuff of the real, workaday world, dominated by the engineer; science, on the other hand, was viewed as blue-sky dreaming undertaken by a band of absent-minded professors. The down-home Thomas Edison, with his short sleeves and cigar, invented useful devices such as the phonograph, the stock ticker, the light bulb; on the other hand, the undoubted genius Albert Einstein, a Jew and a foreigner, dreamed up abstract mind-twisters. Edison was portrayed as a serious worker, like the businessman he tried to become; Einstein, despite his fame and celebrity status, was, like Freud, often the butt of jokes, a caricature of the scientist with frizzy hair and pipe and pacifist beliefs. "Science," wrote the *Saturday Evening Post* in the 1920s, was "the ugly duckling, the elder sister who lives secluded and remote, unknown and unpraised."[8]

Commentators on the fair continually rattled on about science but trotted out technology. Biomedical research, according to the U.S. surgeon general in an essay on the fair, was not a matter of exploring antibiotics or disease mechanisms but of building sanitary hospitals and practicing better public health.[9] The breakthroughs in pharmacology and biology that were occurring—the development of the sulfa and penicillin drugs and the gestation of what is now known as molecular biology—went unmentioned. RCA's charismatic chairman, David Sarnoff, talked enthusiastically about new electronic devices, such as facsimile, but not about breakthroughs in quantum theory that might provide more fundamental advances, such as solid-state transistors.[10] When real scientists did speak, they sounded cautious, dull, even a bit priggish. The University of Chicago's Arthur Compton, a Nobel Prize winner in physics and a powerful figure in research circles, discussed the benefits of science in tones far cooler than those of Wells or Sarnoff. He touched on advances in atomic physics but modestly declared "that it would be years before the significance of these new discoveries is fully known." And he criticized predictions made by planners like Moses and, by extension, exhibits like Futurama. "The most extraordinary and fantastic predictions are made about impending changes in the city," Compton wrote. "I do not believe in these pictures."[11]

To the popular mind of the 1930s, technology and science were indistinguishable. The elevation of prewar technology to a status above science, and the resulting apotheosis of the industrial designer, sprang in part from the fair's desire to attract a large public, a goal it successfully attained. Science was far more difficult to visualize, to embody, than a teardrop car or an elevated highway; it required knowledge and imagination, both rare qualities at any time. As a result, engineers—those dam, bridge, and refinery builders and the leaders of Thorstein Veblen's technocracy who stood so tall before World War II—crowded the stage. There seemed to be little room for the scientists.

The fair thus easily served as an elaborate pedestal for corporate jingoism. Cantankerous Henry Ford, with son Edsel in tow, came to New York to open the Ford exhibit. "Great things will happen this year," Ford snapped at reporters. The fair "is a great thing and it will do the country a lot of good. It will give people something constructive to think about, rather than destructive."[12] Du Pont's vice president for research, Charles Stine, found himself stiffly announcing the development of nylon—a synthetic fabric that would prove more profitable than any other single product in history—before a crowd of three thousand women's club members at a forum sponsored by the New York *Herald Tribune* called "We Enter the World of Tomorrow."[13] The women burst into applause when they thought Stine was introducing stockings as strong as steel. And Sarnoff, an apostle of technological progress, chose the fair as the place to inaugurate America's first electronic commercial television service. On April 30, 1939, RCA broadcast the opening of the fair—Mayor La Guardia ogled the bulky television camera, President Roosevelt said a few words, and Sarnoff gave a rousing speech, proclaiming the launch "of a new industry, based on imagination, on scientific research and accomplishment."

> Now we add radio sight to sound. It is with a feeling of humbleness that I come to the moment announcing the birth in this country of a new art so important in its implications that it is bound to affect all society. It is an art that shines like a torch of hope in a troubled world. It is a creative force which we must learn to utilize for the benefit of all mankind.[14]

Sarnoff then switched to a boxing match. The signals, beamed fifty miles from the Empire State Building, were picked up by less than two hundred RCA sets with eight-by-nine-inch screens, mostly in Manhattan. The service shut down less than a year later, with the FCC accusing RCA of attempting to dominate the new industry with "blitzkrieg" tactics.

## The Process of Maturation

Although big business was undoubtedly stable, it was not frozen in place. Beneath the formidable surface of America's major corporations, pressures were building. Despite the boasting and the breakthroughs, core industrial technologies were maturing. In the post–World War I years, companies had spent increasingly more manpower and money to generate incremental improvements in increasingly complex industrial systems, a telltale sign of maturity. Engineers, particularly in the most advanced fields, drifted steadily toward a more scientific approach and university-based, scientific training; the old rule-of-thumb approaches and the domination of shop training were slowly abandoned as demands mounted. Finally, this maturity revealed itself through a drift toward greater industrial concentrations. As fundamental technological advance decelerated, financial power, generated through volume production, increasingly prevailed, and oligopolies emerged in full force. The automobile companies, for example, could afford to hire masses of engineers and designers to make minor improvements in engines, transmissions, brakes, and lights; to design and implement the alterations in annual model changes (a GM innovation); and to improve efficiencies at their cavernous assembly plants. These costs produced all but impassable barriers to entry for new carmakers. Not only did weaker American nameplates disappear through failure or acquisition, but new domestic competitors were unable to grow large enough to generate competitive economies of scale. Although there were 30 car companies in 1930, the top three—Ford, General Motors, and Chrysler—accounted for 83 percent of new domestic car sales. In varying degrees, such a process of maturation also occurred in the chemical, electrical equipment, and iron, steel, and rubber industries.[15]

This maturation process was exacerbated by economic woes. Having reached geographic limits (international trade was limited by transportation costs and trade barriers), corporations found that the only way to continue to expand was to discover new products or processes or to diversify through acquisition. The notion of economic maturity, articulated by economists such as Alvin Hansen in the 1930s and founded in part on the notion of the end of the frontier, did have some basis in fact. The trend toward corporate diversification that continued so prominently after World War II had actually begun in earnest in the 1920s; by 1939 it had advanced quite far. But to manage a sprawling, diversified, decentralized enterprise, with its multiple cash flows and rigid industrial economies of scale, required sophisticated management systems. As a result, the major firms represented at the New York fair shared a handful of superlatives: the largest companies were engaged in the most complex technological ventures, were the most diversified and decentralized, and had mastered the most sophisticated reporting and control systems. "Diversification," concluded business historian Alfred Chandler in an overview of this scene, "came primarily in those industries which make the greatest use of sophisticated scientific techniques, particularly those developed by modern chemistry and physics."[16]

Despite the wonders of Futurama, two industries—electrical manufacturing and chemicals—represented the apex of technological sophistication in 1939. Electricity had developed earlier—led by Edison, Marconi, Tesla, and Steinmetz and soon to be dominated by Westinghouse, General Electric, AT&T, and RCA—but between the two world wars, chemistry may have provided the most potent economic force. Chemical processes, including petroleum refining, were far more complex than any other industrial system, and the industry engaged in more rapid diversification than any other sector up until 1950. When companies went looking for new markets, they tended to seek out areas that required chemical expertise.

Systematic industrial research began in the chemical industry in 1888, when Du Pont set up a small laboratory to investigate smokeless powder; in 1902 it organized its more ambitious Eastern Laboratory to investigate dyes and organic chemicals. Between 1899 and 1946, fully a quarter of all industrial laboratories set up were involved in chemistry. Before World War II, no industry had a higher proportion of scientists and engineers than chemicals.[17] Unlike other basic sciences, such as physics and biology, the gap between chemical discovery and chemical engineering had narrowed, accelerating the pace of innovation in both research and development, particularly earlier in the century.

One tangible sign that chemistry had come of age as a technological force was the development, around the turn of the century, of chemical engineering as a discipline. The Massachusetts Institute of Technology, MIT, was then a fairly small, rather ordinary technical school for undergraduates along the Boston and Albany railroad tracks in Back Bay Boston. A chemistry department had been established in 1888, but the major shift in priorities occurred in 1904, when William H. Walker, a partner of pioneering chemical consultant Arthur D. Little, returned to MIT to teach.[18] Walker began to preach the merits of industrial–academic ties and created a new curriculum that blended industrial concerns with state-of-the-art chemistry. In 1905 he started a laboratory that would perform, for a fee, research for chemical firms, particularly those that lacked research facilities. Walker thus created an intimacy with industry that shaped the school. (Around the same time, Prof. Dugald Jackson was building an electrical engineering curriculum along similar lines.) In 1916, Little, Walker's former partner, urged the MIT corporation to establish the School of Chemical Engineering Practice, which opened its doors in 1920. The school, directed by Walker and funded by George Eastman (of photography fame), was intimately concerned with industrial problems, particularly the efficient, continuous mass production of chemicals. "Chemical engineering has enabled the businessman and investor to view the chemical reaction from the standpoint of efficient and economical manufacture of a product and to accurately forecast the profits to be expected from the application of a chemical reaction," Du Pont's Stine wrote in 1928. He summed it up more epigrammatically: "Chemistry views the chemical reaction; chemical engineering views the pocketbook reaction."

Early on, Walker established powerful ties to industry, which gave him enormous political power within the MIT corporation. When MIT moved to Cambridge in 1916, the money for handsome new quarters along the Charles River was provided not by the school's earlier Brahmin benefactors but by chemical industrialists and Walker supporters—notably, the du Pont family and Eastman. A year later, Walker was made director of the Division of Industrial Cooperation and Research in an attempt to formalize across the university what Walker and Dugald Jackson had already accomplished in their own areas.

By 1930, however, the pendulum was swinging back toward basic research. In an uncertain atmosphere of financial duress, the MIT trustees appointed Karl Compton, a physicist from Princeton and Arthur Compton's brother, as president. MIT's Research Laboratory of Applied Chemistry (RLAC) had increasingly focused on narrow industrial concerns—particularly as more companies set up their own industrial laboratories. With the advent of the depression, industrial support dried up. Meanwhile, Compton had secured a large grant from the Rockefeller Foundation for science-based education and aggressively moved to build up the science departments, including chemistry. Compton did not eliminate corporate sponsorship, nor did he drive the RLAC out of business, but he put it on a new basis, forbidding applied research that could be performed by consultants such as Little. By the end of the decade, the school had moved away from Walker's singleminded focus on industrial ties and back toward science-based training and research.

The situation at MIT mirrored larger changes in the relationship of science and technology to commerce. As chemistry blossomed, engineers could produce large gains from new products. But as the industry matured, as chemical engineering knowledge diffused, companies had to mine deeper veins of science. That drift toward basic science also occurred at Du Pont, both the largest and most advanced company in the American chemical industry. In the early decades of the century, Du Pont had been reorganized by three du Pont cousins—Alfred, Coleman, and Pierre—each of whom had been trained at MIT (though prior to the establishment of Walker's chemical engineering curriculum). Over the next few decades, Du Pont transformed itself from a trust of horizontally linked units into a single, vertically integrated corporation.[19] Perhaps more significant was that Pierre and Coleman built upon the work of

Frederick Taylor, the pioneer of scientific management, to develop more sophisticated methods of tracking costs and returns. Pierre du Pont in particular encouraged the development of sophisticated cost accounting measures that allowed management to generate a quantitative sense not only of profits but of the return on invested capital. The company also set up separate units to specialize in such tasks as designing and building new plants and in conducting research. It was certainly no coincidence that Du Pont's attempt to capture the dynamic process of financial change closely resembled the monitoring and analysis of chemical processes of Walker's chemical engineers.

Du Pont, like most of its industrial brethren, was built upon economies of scale. Was there a link between chemistry and those economies, between a particular scientific base at a specific stage and a corporate structure? Chemical processes clearly demanded considerable capital; no one would bother to spend the necessarily large sums on minor products. Thus, if Du Pont had thought that nylon was suitable only for very limited, niche markets, it would not have spent millions of dollars and thousands of man-hours laboring to develop the intricate commercial processes required to scale production up. In fact, the production process for nylon forced Du Pont to call a variety of skills available only to a large enterprise—from the efficient synthesis of essential intermediates, to the spinning of nylon fibers, to the knitting and dying of finished materials. Scale economics, however, also requires a minimum amount of technological stability, which chemistry, despite its power and promise, provided. Du Pont was able to wring substantial returns from its major new products of the 1930s because no other corporation was able to leapfrog it technically. Its patent position on nylon was particularly impregnable, and the company was able to spend five years scaling up nylon, without a hint of a competitive threat (thus further strengthening its hold over the market). The pace of progress in chemistry thus favored the large corporation, as it did in electrical equipment, steel, and automobiles.

Du Pont also realized early on how further exploitation of chemistry required more extensive harvesting of basic science. In the years after World War I, the company had diversified away from a dependence on munitions by copying dyes patented by German firms, making agreements for acquiring patents or products, and acquiring new operations. The company's major emphasis, common in that era, was on patents. "The process of invention, management believed, was too unpredictable to institutionalize within the corporate structure. So Du Pont kept a sharp watch for new developments anywhere in the world but especially in small entrepreneurial companies."[20] In 1926 Stine began to argue before the company's executive committee for a new approach. He sought funds for a unit designed to operate in parallel with normal, product-oriented research, thus imitating the systematic performance of both "R" and "D" that the Germans had pioneered and that AT&T and General Electric had already instituted in America. Stine, citing a speech given by Secretary of Commerce Herbert Hoover calling for greater private investment in pure research, warned that "applied research is facing a shortage of its principle raw materials." In later memos, Stine carefully spelled out exactly what he was looking for. He distinguished between fundamental research—he avoided using the term "pure science" so as not to scare off the executive committee—and scientifically applied research. The latter, he said, normally had a practical goal, which often could be achieved. Fundamental research, on the other hand, as the historians of Du Pont research paraphrase bluntly, was "a form of gambling." Stine argued that, in the long run, fundamental research would allow "Du Pont to treat materials in a more rational and scientific manner."

Stine had a wish list of areas he wanted to explore: colloid chemistry, catalysis, organic synthesis, polymerization. Each was at the forefront of scientific chemistry. The year 1926, the high noon of the New Era, turned out to be highly profitable for Du Pont and remarkable for its largest single investment, General Motors (nearly three-quarters of Du Pont's net income came from General Motors dividends), which was newly reorganized by Alfred Sloan, an MIT-trained electrical engineer. The executive committee not only lavished on Stine a budget of $300,000 a year, much more than he had sought, but urged other product units to suggest new areas the company should seek out. Stine immediately built a new laboratory, dubbed Purity Hall, and went looking for some twenty-five "pure" researchers.

Stine's timing was fortuitous. With the coming of the depression and an upsurge in antitrust suits against Du Pont and other large corporations, the company backed away from the active program of external diversification through acquisition. And, of course, Stine's efforts, particularly in the polymer area, were spectacularly successful. Rayon, neoprene, and nylon all stemmed directly from the basic research of one man, Wallace Carothers, a Harvard chemist whom Stine had recruited. With the fruits of Purity Hall, Du Pont remade itself again in the 1930s, completing its retreat from the maturing gunpowder business, reducing its acquisition efforts, and plunging into the new world of internally invented synthetic fabrics and other, advanced, chemical products.

Science, both fundamental and applied, provided the ultimate refuge from antitrust attack and the New Deal's oft-expressed (though rarely acted upon) desire to nurture smaller businesses. Despite those pressures, however, there were few signs in the late 1930s that the hegemony of the large, decentralized corporation such as Du Pont was ending. In fact, the creativity of Du Pont seemed to suggest just the opposite: given the financial deficiencies of academia and of smaller businesses, and the refusal of the government to fund widespread research, the domination of both science and technology by the large corporation seemed to be ensured.

And yet, even as the New York fair opened, the world was changing in fundamental ways. Europe was about to go to war. By the time the fair closed in October 1939, Germany had invaded Poland and turned its mechanized forces to the west. By the time 1940 dawned, there was already something anachronistic about the World of Tomorrow.

## Atomic Scale and Scope

World War II, more so than other wars, proved a powerful incubator of technological interaction and serendipity. Old barriers fell; disciplines that had never mixed were suddenly jostled together like troops on a crowded transport. Pure mathematicians met atomic physicists who mixed with biologists, chemists, metallurgists, even economists and social scientists. Scientists who never considered the rigors of utility found themselves assuming the roles of engineers; and engineers who eschewed abstraction discovered that they had to master higher mathematics and gain a better idea of quantum physics in order to work with scientists. The result was an astounding array of new devices, new technologies, new ideas: from famous products such as the digital computer, the jet aircraft, penicillin, radar, rocketry, and nuclear power (and nuclear weapons), to a plethora of more obscure, if significant, advances in production techniques and diverse, increasingly sophisticated instrumentation.

The most obvious lesson of the war is that of the power of enormous scale. The Manhattan Project was far larger and far more sophisticated than any R&D project ever attempted—and it was concentrated over a short time period, on an extremely focused goal: a handful of workable bombs. The entire effort cost $2 billion in 1945 dollars. At the time, given the fear that Germany or Japan would develop a bomb first, the few people who were privy to such secrets did not worry much about cost (in the light of later federal projects, the amount no longer seems so staggering).

But a budget of $2 billion was far greater than the R&D budgets of even the largest and most scientifically oriented companies. In 1939 General Electric, America's second largest corporation and one of the country's most scientifically sophisticated, generated only $304.7 million in total sales and spent about 4 percent or so, or $12 million, on R&D. And Du Pont, after a decade of remarkable productivity, spent only $10.45 million on ten major areas of research interest on $299 million in sales.

The Manhattan Project blurred the line between Stine's notions of fundamental science and applied science. On the one hand, there was certainly a practical goal; on the other, it involved the participation of "pure" scientists, including theoreticians and mathematicians, in a mission that was, to say the least, problematic. In 1941 all that existed of an atom bomb initiative was a handful of physicists from a variety of countries working in New York, Chicago, and San Francisco and one crude atomic pile going up beneath the University of Chicago's abandoned football stadium.[21] Four years later there existed not only a usable bomb but a secret empire of laboratories and processing plants, including a sprawling R&D facility—a research city that hardly resembled the sleek dreams of Bel Geddes—atop a New Mexican mesa called Los Alamos. To make the bomb work required enriching uranium-238 to produce fissionable material, either uranium-235 or plutonium. And that process required extremely complex and capital-intensive separation or enrichment systems. Thus, although the rapid construction from scratch of Los Alamos remains the most famous use of the government's money, the construction of Hanford and Oak Ridge, the two centers for uranium processing, remain the most awe inspiring and cast the strongest illumination on a major theme of postwar technology: the rise of the giant systems projects.

At Oak Ridge, Gen. Leslie Groves, the military head of the project who had built the Pentagon a few years earlier, bought fifty-nine thousand acres of impoverished Tennessee hill country. Groves planned to build at Oak Ridge the first of three experimental processes for producing uranium-238. To provide transportation, he built fifty-five miles of railroad and three hundred miles of paved roads and streets. A town was built (designed by the Chicago firm of Skidmore, Owings & Merrill, later famous for its corporate headquarters) for thirteen thousand people.

At the heart of the process was a massive scaleup of the calutron, a machine that separated out materials of slightly higher electrical charge by spinning them about a magnetic "race course." The calutron was a sibling of the cyclotron, invented by the University of California's Ernest O. Lawrence and M. Stanley Livingston and a key tool for probing the atomic nucleus. In 1939 Lawrence won a Nobel Prize for his cyclotron work and had assembled a considerable power base within the national research establishment. He was the major voice pushing the calutron option and contracted out much of the work to five classic World of Tomorrow corporations—Eastman Kodak, General Electric, Allis-Chalmers, Westinghouse, and Stone & Webster—"the most spectacular concentration of industrial power this country had yet seen."[22] In 1943, construction of the actual "manufacturing" facilities began: Stone & Webster, one of the largest construction companies in America, started to pour foundations for the so-called Y-12 facilities even before basic decisions about the number and configuration of the calutrons had been made. Twenty thousand workers scrambled to build the facilities; forty-eight hundred workers were trained to operate the untested machines—and had to be paid, housed, and distracted when delays shut down the works. Many of the systems going into the massive plant pressed techno-logical limits. Calutrons themselves had never before been built to such a size or run at such length. No one had time to build either a pilot plant or an intermediate plant.[23] The factories themselves were the longest in America; workers had to use bicycles to patrol them. In 1942 Groves thought electromagnetic diffusion might cost $12–$17 million, a figure that was then increased to $35 million. In 1946, Groves discovered that construction costs alone amounted to $304 million, with another $204 million allotted to operating costs.[24] The scale was frightening:

> Alpha and Beta buildings alone eventually covered more area. . . than twenty football fields. Racetracks were mounted on second floors; first floors held monumental pumps to exhaust the calutrons to high vac-uum, more cubic feet of vacuum than the combined total volume pumped down everywhere else on earth at the time. Eventually the Y-12 complex counted 286 permanent buildings large and small—the calutron structures. . . chemistry laboratories, distilled water plant, sewage treatment plant, pump houses, a shop, a service station, ware-houses, cafeterias, gate houses, change houses and locker rooms, a paymaster's office, a foundry, a generator building, eight electric sub-stations, nineteen water cooling towers—for an output measured in the best of times in grams a day.[25]

And yet Oak Ridge's Y-12 almost went off the rails. There were recur-rent vacuum leaks. The huge electromagnets that made the calutrons work began to creep across the floor—they were finally strapped down with steel belts—then began to short circuit. This was a major disaster, if only because the silver that made up the coils (copper was scarce because of other war needs) had been borrowed from the Treasury

Department, forty-seven thousand tons worth $300 million. Despite constant repairs, the magnets had to be shipped back to Allis-Chalmers in Milwaukee to be rebuilt—a grievous expense and a major delay. And for all that, Y-12 proved a relative failure. While the calutrons did produce some usable bomb material for Los Alamos and absorbed the greatest chunk by far of the final bill, they proved to be the least successful processing method. Lawrence's calutron, in fact, became the butt of jokes among some of the physicists during the war and is today obsolete.[26] But while the Hanford reactors and the gas-diffusion facilities at Oak Ridge proved far more efficient in providing bomb material, Y-12 arguably made the bomb possible during the war, albeit at tremendous cost.

Such are the possibilities of an all but unlimited checkbook. There was not necessarily any theoretical reason why a bomb could not have been built at considerably less cost, it just would have taken longer. Capital's ability to bring vast resources into play on a problem was focused on one goal: to accelerate the speed of development. Capital was used to build vast, redundant systems, some of which were never perfected before the war had ended. It was used to follow parallel paths instead of pursuing one at a time, with the hope that one would pan out. And it was used to begin building plants and infrastructure before scaleup and testing had been completed. Capital at this scale allowed great risks to be taken, major mistakes to be made. But scale capital also brought with it irritations and difficulties. Many of the scientists felt lost or harassed; others, such as Hungarian physicist Leo Szilard, chafed under military authority and hated the requirements of secrecy and bureaucracy (and thought them counterproductive). The enormous and inevitable waste would have made the project laughable as a commercial venture. Although the Manhattan Project became the popular model for technological achievement after the war, it could be accomplished only by one patron, the federal government, which operated outside the markets, outside the bounds of ordinary peacetime necessities.

Ironically, with the advent of undertakings such as the Manhattan Project, the large corporations would discover just how expensive science really could get and, for the first time, would themselves feel the sting of exclusion and dependency. Despite their large earnings, which swelled with war work, and their ability to raise vast sums on Wall Street, the large corporations could not meet the capital demands of certain scientifically oriented war projects without the financial help of the government. In a sense, the large corporations found themselves squeezed out, much as they had squeezed out the lone inventor and the smaller corporation earlier in the century.

## The Radar Model

The Manhattan Project was the model for a certain kind of systems project characteristic of the postwar era: the space program, the construction of high-energy accelerators (including the successors to Lawrence's Berkeley lab, even the proposed, now dead, Superconducting Supercollider), the Strategic Defense Initiative, the weapons development programs of the military–industrial complex itself. But in terms of important technological fallout and a civilian style of technological development, the Manhattan Project pales in comparison with another major government R&D effort centered not on an isolated mesa in New Mexico but in the busy heart of Cambridge, Massachusetts.

The American effort to develop and perfect microwave radar was, in some ways, a more traditional research project than the Manhattan Project, albeit on a war footing. It linked a century of earlier investigations into the nature of electricity—going back to Michael Faraday, James Clerk Maxwell, and Heinrich Hertz—to the postwar explosion of solid-state electronics. Although the radar project was both sophisticated and successful, it involved the improvement of an existing system rather than the invention of something brand new; Stine would have easily called it applied research. Unlike the bomb project, the radar group developed not just some workable prototypes but multiple systems that could be manufactured, installed, and operated in all kinds of conditions. By 1945, some 150 different radar systems had been made. While the direct progeny of the Manhattan Project—the atomic energy program and weapons manufacture—never escaped the clutch of the government, the major offspring of radar, the semiconductor, defines the most fertile and productive postwar commercial technologies. The radar project even consumed more cash—$2.5–3 billion[27]—than the atomic bomb program, in part because it began much sooner, in 1940, before Pearl Harbor. "The bomb ended the war," said Alfred du Bridge, who ran the government's primary radar research facility. "The radar won the war."[28]

In the 1930s, America had clearly fallen behind in radar. But as war broke out in Europe, the same small group of research administrators who ran the federal National Research Defense Council (NRDC; the oversight agency for government R&D and the predecessor to the wartime OSRD)—Vannevar Bush, Karl and Arthur Compton, Alfred Loomis, and Ernest Lawrence—quietly decided that research into radar was a priority (the atom bomb was not yet on the agenda). Loomis, the cousin and financial advisor of Secretary of War Henry Stimson, served as the facilitator. Born in 1887, Loomis had attended Andover, Yale, and

Harvard Law, enjoyed a successful law practice and a fabulous investment career (at Bonbright & Co. he made a killing on 1920s utility bonds, then got out before the crash), sponsored America's Cup yachts, rode and hunted on his Hilton Head Island preserve, and made himself into a physicist after mastering ballistics during World War I—"the last of the great amateur scientists," according to Berkeley physicist Luis Alvarez. In the late 1930s Loomis developed a fascination for Lawrence's Berkeley operation and helped to raise funds for Lawrence's cyclotrons (he also raised money for physicist Enrico Fermi's first experimental pile at Columbia University). His real passion, however, was radar. Loomis had set up a radar laboratory at his stone mansion in Tuxedo Park, New York, and his old friend Karl Compton had named him head of the Microwave Committee of the NRDC.[29] Although the country was not yet in the war, America had the resources and skills to press forward with certain military technologies such as radar. The British, on the other hand, had advanced a number of vital military technologies, particularly radar, quite far, but lacked the resources (and, faced with the German might, the time) to go much further.

In early 1940 the barriers between U.S. and British technology establishments began to fall. In February, the British sent over its new Rolls-Royce Merlin engine, which was powering its Spitfire and Hurricane airplanes (the Merlin would transform the American P-51 Mustang into an important long-range fighter). The British also quietly passed along the calculations of Otto Frisch, a refugee from Hitler working in the British atom project, which showed that a bomb would require pounds, not tons, of U-235 to function. Soon afterward, Fermi began assembling a team in Chicago, under the administration of Arthur Compton, to attempt a self-sustaining chain reaction.[30]

But the most important technical exports came in a black box carried to America by the British cruiser *Duke of Richmond* in August 1940. That box was the payload of a mission headed by Sir Henry Tizard, an Oxford chemist who had been organizing and advocating a scientific role in military affairs since 1934 and had convinced Scottish physicist Robert Watson-Watt to come forward with his early radar work. Tizard's mission was specifically to seek American government help in developing military technologies. As incentive, he brought plans for a power- driven turret for America's Flying Fortress (which would become the dominant long-range bomber in the war), news of something called jet propulsion, and an assortment of gadgets such as the proximity fuze and the Bofors repeating cannon with predictor gun sight. He carried the most important device in a small black suitcase: a simple-looking vacuum tube developed by two

Birmingham scientists and manufactured by Britain's General Electric Corp. This ten-centimeter-cavity magnetron was capable of putting out extremely intense, shortwave radiation for use in radar systems.

The key to the radar program was of course electromagnetic radiation, or radio waves. Radar used electromagnetic radiation to identify distant objects, such as enemy boats or ships (the term *radar,* meaning radio detection and ranging, was coined by the U.S. Navy in 1942). Watson-Watt's early radar system was simple, at least in theory. A transmitter would send out a pulse of electromagnetic radiation into space. Those waves would strike a distant object—say, the metallic hull of a Luftwaffe bomber—and a much smaller wave would ripple back, or echo; that wave would then be detected, measured, and recorded before another pulse would be sent out. By knowing the finite speed of light and measuring precisely the time between the pulse and the echo, a radar system could plot distant objects.

By the 1930s, radar work in Britain, Germany, and the Soviet Union had proceeded quite far despite major technical limitations. As electromagnetic waves are beamed out, they tend to broaden and weaken; the lower the frequency (or the wider the wave), the bigger the problem. To generate a stronger, tighter, higher frequency beam would require either a higher-frequency vacuum tube transmitter or a larger dish. The British, in the early days of the Battle of Britain, had set up their equipment with large antennas atop wooden platforms scattered, like the remains of electronic Druids, along the cliffs of Dover. Such large dishes, linked to Royal Air Force aerodromes, succeeded in detecting waves of German bombers droning across the channel at night (this so-called Home Chain system was instrumental in winning the Battle of Britain), but they were too large for use by planes or ships. They could not, for instance, detect German U-boats prowling among shipping lanes or a single fighter plane approaching from several miles out. The task then was to develop a powerful transmission source that employed beams with wavelengths well into the microwave region—tight, short waves that diffused slowly.

The key component of radar was the so-called vacuum tube, or valve. A sophisticated cousin to the light bulb, the vacuum tube consisted of a heated filament and a metal plate encased within a vacuum-sealed glass bulb. By sending current through the filament and heating it, electrons could be produced that streamed toward the metal plate. Early in the century, various inventors, from Edison to Lee de Forest, discovered that the stream of energy could be manipulated for a variety of purposes: rectification (transforming an alternating current into a direct current, which is necessary for radios, televisions, and telephones), amplification (using the energy flowing off the filament to amplify a weaker, external controlling current—say, a radio signal—flowing into a so-called grid), detection, and oscillation (signal transmission). By the 1930s, tubes were being used to transmit signals and, replacing the crystals in ordinary radios, receive and amplify them. The tube business boomed. Between 1936 and 1940, some one hundred million vacuum tubes were sold annually, at a sixth of the price they were being sold for in 1922.

Tubes had their deficiencies—they were fragile and bulky, consumed large amounts of power and radiated lots of heat, and, like any light bulb, would eventually burn out—but they were fairly well understood. And a steady stream of improvements had created a range of specialized tubes, lowered their cost, reduced their size, and considerably boosted their durability.

The magnetron was a specialized version of the vacuum tube—a power tube, in the electronics jargon—designed specifically to transmit strong pulses of microwave radiation. An early magnetron resembled a small hubcap with nine holes punched in it—a large one at the center and eight around the circumference. The eight rim holes, or cavities, were connected to the large center hole by smaller openings. When direct current was pumped into the cathode in the center of the magnetron, it sent a stream of electrons hurtling into the outer cavities and across a magnetic field that had been set up. The cavities caused the electrons to resonate, generating extremely powerful, microwave pulses.

The appearance of the cavity magnetron at Tuxedo Park changed the direction of American radar efforts. The NRDC, seeking a facility with technical assets near a large body of water, proposed to form a laboratory to pursue radar. The group, dominated by MIT alumni and corporation members, chose the engineering school in Cambridge. Loomis agreed to equip the lab with the necessary hardware; Lawrence dashed off to find staff. Tizard's team had suggested using young physicists, mostly from academia, familiar with pulsed electromagnetic waves, rather than engineers, mostly from industry, who were more comfortable with continuous, analogue waves used in radios and televisions. This suited Lawrence just fine. Quickly, he recruited du Bridge, the young head of the physics department at the University of Rochester, to run the new laboratory. Then the pair began hiring the cream of the crop of American academic physicists, including Columbia's Isidor Rabi (who had been experimenting with molecular beams), Alvarez, and Edward McMillan from Berkeley. Those early recruits, in turn, made further contacts. The Radiation Laboratory began with twenty people in Room 4-133 at MIT; at the height of its life it employed thirty-eight hundred people, of which a thousand were academic scientists—one in five were American physicists. Its manpower was double that of MIT in peacetime, and its last budget exceeded MIT's endowment.[31] Similar, if smaller, labs were also established at Columbia and Harvard. Bell Laboratories pursued an intensive program of radar research as well.

The Radiation Lab took on its unique character not only from this preponderance of scientists—in a 1945 article, *Fortune* insisted on calling them "longhairs"[32]—but because of its catholic approach to its mission. Although the lab was dominated by physicists, du Bridge attracted people from other scientific disciplines as well; even Paul Samuelson, the future Nobel Prize–winning economist, applied his mathematical skills at the Rad Lab. Generally, this was a group comfortable with the impressive, if recondite, structure of quantum mechanics, a field that also formed the background of the Manhattan Project. Quantum physics was relatively new, powerful, and extremely abstract. The phenomenon at the heart of both projects made logical sense mathematically, but, like the notion of electrons as both particle and wave or the seeming paradox of Heisenberg's uncertainty principle, were counterintuitive when imagined mechanically, as a good Newtonian might. Though many of the Rad Lab team were skilled experimentalists, their modus operandi was not empirical but theoretical: that is, they labored to understand how things work. Science historian Daniel Kevles captured the mood of the group in an exchange soon after the English magnetron arrived in Cambridge: A group of theoretical physicists stood around a dissembled magnetron scattered across a table. "How does it work?" someone asked. "It's simple," said Rabi. "It's just a kind of whistle." Edward Conden, one of the physicists, responded, "Okay, Rabi, how does a whistle work?"[33]

Perhaps more remarkable was that the Rad Lab succeeded in extending this willingness to establish interdisciplinary links to its ultimate "customers," the various military services. Du Bridge later argued that the Rad Lab's success was due in part to the NRDC decision not to place it under military control, as it would the Manhattan Project. Instead, the Rad Lab acted as a civilian contractor, negotiating with the various military interests at arm's length. After the war, du Bridge described civilian control as a beneficial form of "corporate irresponsibility," adding that "many of the most effective weapons to come from Radiation Laboratory developments were quite reasonably regarded as misguided when they were in their earliest laboratory stages. A service laboratory might have regarded speculative research as a luxury which it could not allow itself."[34] If the secrecy and military flavor of the Manhattan Project became the model for postwar weapons development and nuclear research, the Rad Lab, with its academic flavor and contract system, became the model for postwar civilian R&D funded by the government.

This crowd of physicists proved to be enormously productive, a shock at first to the military authorities. The Rad Lab developed a variety of radar components—more powerful magnetrons; waveguides; detectors, called T/R boxes, for separating weak, incoming signals from radio noise; new means of measuring tiny intervals of time; new circuits and phosphors for cathode-ray tubes—and then proved surprisingly capable of effectively dealing with the engineers at manufacturing firms such as Westinghouse, RCA, Philco, Raytheon, Western Electric, and General Electric who fabricated, mass produced, and often fed back suggestions for improving the devices. Soon, U.S. industry was churning out two thousand radar sets a month. The American electronics industry swelled from 110,000 employees before the war to 560,000 in 1941; sales grew twelve times.[35] By the end of the war, Allied troops were using one-centimeter radar in ships and planes to locate submarines—the German U-boat threat had long abated—to lay down gunfire and, with Loran, to master accurate, long-range navigation. And yet, for all their understanding of magnetrons and waveguides, based on the quantum model, they discovered that they needed one component that still eluded their comprehension: the humble crystal.

Crystals for radio-wave detection had been around as long as vacuum tubes. The first crystal was discovered in 1886 by German chemist Clemens Winkler; in that passionate nationalistic age he called it germanium. Crystals were odd, discomfiting. Under certain conditions they could suddenly change from electronic insulators to electronic conductors—thus they became known as semiconductors. Unlike metals, they were made of a myriad of regular molecular shapes, fitting together at the molecular level like a vast, jagged jigsaw puzzle. Unlike insulators, their conductivity seemed to increase as they were heated. When they were exposed to a magnetic field, they reacted in unexpected ways. Tiny impurities, particularly along the surface, or interface, caused performance to vary widely. And it proved difficult to fabricate them with consistent purity. Crystals were also notoriously difficult to work with. To tune them, early radio operators had to make careful adjustments on a thin, springy "S" wire contacting the surface—the once-famous cat's whisker. Nonetheless, they worked, though they were gradually replaced in most ordinary radios by tubes. Then came radar, with its much more rigorous demands. Radio tubes proved unable to detect radar's very weak incoming echo with its rapid oscillations (electrons moved through a tube at the same rate as a microwave cycle); crystals, however, could. And so, despite their deficiencies, semiconductor crystals of silicon and germanium found themselves in radar devices.

But how did they work? The physicists at the Rad Lab had only the most cursory of explanations as to how a crystal functioned at the quantum mechanical level. In the 1930s, interest was revived in investigating the phenomenon, and Walter Schottky, a physicist at Siemens in Germany, had suggested that the key to semiconductor behavior might lie along the interface between needle and crystal. But what was taking place along that interface? No one really knew. Researchers during the war did attempt to learn more about impurities by adding atoms of various elements to semi-conductors, but it was working in the dark. And so, while researchers understood the actions of electrons in a vacuum tube, the crystalline device remained the province of the true Baconian empiricists. They added impurities. They worked with a variety of different kinds of crystals. They experimented with production techniques.[36] They recorded the results. Was there a pattern? Did it suggest a mechanism?

In the end, radar, the most sophisticated electronic device used during World War II, involved both semiconductor crystals and tubes. The devices functioned well, though the crystals still proved delicate and difficult to work with, not to mention that they were still resistant to theoretical understanding. To an outsider, the radar systems seemed to be four-fifths buzzing high technology, based upon sophisticated quantum physics—that is, the tubes—and one-fifth black box—the mysterious crystal. How shocking then that little more than a decade after the war, the tube would be a dying technology, and the crystal would be transforming the electronic landscape.

# 3

## Optimists and Pragmatists

### The Potency of Science

November 2, 1945: General MacArthur meets Emperor Hirohito on the steel deck of the battleship *New Jersey;* the war has finally ended. With the shock of, in the emperor's own words, "those new and cruel bombs" passing, physicists such as Robert Oppenheimer and Enrico Fermi found themselves celebrities, lionized by an adoring press and public, sought out by politicians. "I have become convinced, and I think many of my colleagues are convinced, that the free and unrestricted development of basic scientific research is such an important part of our potential natural resource that we cannot afford to neglect it," Sen. Harley Kilgore of West Virginia told a scientific group in 1945.[1] In 1948 *Fortune* described Oppenheimer as "the youngest, most brilliant brain to emerge from the atom-bomb project," then went on to quote approvingly a passage from an essay by physicist Edward Conden: "Society is at this moment at the threshold of an undreamed-of-mastery of our material environment, for science, which provides that mastery, is in its Golden Age."[2]

The rigorous, quantitative, interdisciplinary style of wartime physics made a deep impression on the shape of postwar R&D. The momentum of wartime science was palpable, and it produced startling new vistas and surprising cross-fertilizations. In Cambridge, economists led by Rad Lab veteran and MIT professor Paul Samuelson labored to quantify the somewhat gnomic utterances of John Maynard Keynes in his *General Theory.* In Chicago, a group that included at one time or another as many as nine future Nobel Prize–winning economists met at the Cowles Commission to discuss everything from ways of modeling the macro-economy (a field soon to be known as econometrics) to how people make decisions, economic and otherwise. At Cold Spring Harbor on Long Island, physicists Max Delbrück and Leo Szilard huddled with

biologists to wrestle with a new molecular approach to biology. In Santa
Monica, California, the newly formed Rand Corporation assembled physicists, mathematicians, and social scientists to brainstorm everything from
safe nuclear reactors to new kinds of nuclear weapons and defenses. In
Detroit, a group of bright young army men—the whiz kids—led by
Robert McNamara settled in to manage Ford Motor in a more quantitative fashion. And in New York City, a diverse group of scientists, including
anthropologists (Gregory Bateson and Margaret Mead), psychologists,
neurophysiologists, and mathematicians (John von Neumann and Norbert
Weiner), met regularly under the aegis of the Josiah Macy Jr. Foundation
to sketch the outlines of a new field soon to be called cybernetics.[3]

These meetings were emblematic of a new interdisciplinary spirit.
The strict division between science and technology, which only the
largest companies could transcend before the war, gave way; so, too, did
the line between social science and hard science. The rising tide of federal money into research, mostly from the military in the 1950s, encouraged not just an inclusiveness of subject matter but affiliation, a sort of
research pluralism: the walls between the government, the military, the
nonprofit community, business, and academia broke down.

It was as if a new, subversive force had been released—science—and
no one knew quite how to channel it. Would the federal government
ensure that the well of basic research, tapped during the war, be refilled
during the peace? Would the federal government play a major role in
postwar science and technology? And if it did, how would that role be
structured? Would corporations continue to pursue war-related businesses, thus remaining dependent on government funding? Would they
return to the prewar emphasis on product development, engineering, and
technology over science, or would they plunge into basic research and
thus wrestle with the difficult question of science's relationship with product development? Which scientific developments would lead to commercial development? And which could the corporations afford to pursue?

How these questions were answered involved a complex array of
individual factors, beliefs, prejudices, and politics. Uncertainty ruled.
No one knew what kind of world faced the victors in the war. From
1940 to 1944, corporate profits had swelled by 60 percent. But Europe
and Japan had been devastated, and there were loudly proclaimed
predictions that America itself would now suffer through a severe
inflation or savage recession—perhaps even the return of the Great
Depression. When the peace broke out, military budgets were
slashed, and the boys, seeking jobs, flocked home. Others still spoke of
the fear, dating from the economic recession of 1938, that America

had become a mature economy, consigned in peacetime to slow growth and naggingly high unemployment. Moreover, individual companies had their own skills, their own predilections to consider. Relatively few firms had participated in the most advanced war work. Thus, for the majority of companies, plans were laid for a return to business as usual after the war, to a world without shortages, rations, or government red tape.

This new environment would produce two different visions of how to manage highly charged, scientifically driven, technological change— two visions that would reappear in a variety of guises throughout the postwar period. The first, the optimistic view, envisions technology transcending everyday limitations. The corporation, or the society, resembles a fresh canvas; only the future matters, not the past. Accounting, management, budgeting, and forecasting—or making distinctions between science and technology and between the social and physical sciences— were exercises that technologically stimulated growth would quickly render obsolete. Such exercises were like buying clothes for a fast-growing youngster: who cares how we look today, if tomorrow we will be so very different?

The second view, the pragmatic approach, recognizes the power of science and technology but seeks to fit it into structures that already exist. Technological possibilities were not larger than the already established business itself; transformation was not in the works. Technology had to be dammed and channeled, not released to wander through the woods. Different businesses, different technologies possessed different intrinsic possibilities and limitations. Corporations had pasts, they had current responsibilities to labor and shareholders. They had, in many cases, legacies to preserve; not all social and economic problems could be solved through technological means. Success and prosperity required discriminating choice.

The immediate postwar period was electric with possibility. Like the seminal scientific gatherings after the war, a small group of corporate executives and government policymakers wrestled with issues involving science and technology that had never been raised quite this way before. The technological world in 1946 was, in a very real sense, a new world, and it resembled 1993 far more than it did 1939. The decisions and attitudes of key business managers and policymakers thus had an elemental quality, like the first roads laid across a newly discovered continent.

## Raytheon's Entrepreneurial Dreams

Until the mid-1950s, the Raytheon Manufacturing Co. was a firm engaged in a constant round of costume changes.[4] Even the story of its founding is replete with ironic asides, coincidences, and sudden turns of fortune. Shortly after World War I, Vannevar Bush, a young electrical engineer at MIT, signed up as a consultant for a Massachusetts electronics company called American Research and Development, or AMRAD, which was bankrolled by J.P. Morgan & Co. in New York. One day an AMRAD engineer named J. L. "Al" Spencer showed Bush a flatiron thermostat he had invented at home. Bush—young, energetic, and ambitious—promptly recruited an old Tufts roommate named Laurence Kennedy Marshall to organize a company to manufacture and market Spencer's thermostat. With the defection of Bush and Spencer, AMRAD promptly folded and Morgan ended up with all its patents, while absorbing a good-sized loss. At that point, an out-of-work AMRAD engineer named Charles Smith approached the trio about one of his inventions, a novel refrigerator. Bush and Marshall promptly formed a second company, the American Appliance Co., to develop it.

Alas, the thermostat fizzled, the refrigerator flopped, and another company began to complain that it already owned the name American Appliance. Fortunately, Smith had also invented a radio tube, called the Raytheon. Marshall managed to convince Morgan to sell him the patent rights to the tube—for $50,000, mostly in American Appliance stock—and renamed the company Raytheon Manufacturing. With Morgan as a big shareholder, Raytheon prospered in the tube business, particularly through wide contacts into the academic research world provided by Bush. Through Bush, Raytheon retained close ties to the MIT establishment and later, as Bush moved to the Carnegie Institution and then into the government, to the higher reaches of the federal research community.

If Bush was an expert guide to the outside world, the defining force behind Raytheon was Marshall, a physicist-turned-construction-engineer-turned-entrepreneur who provided great drafts of enthusiasm and vision for the company. "[At Tufts] I think I was probably one up on Marshall in regard to mathematics," wrote Bush in his memoirs, "but he was certainly several jumps ahead of me in understanding the kind of world we proposed to enter and challenge."[5] As evidenced by the quick shift from thermostats to refrigerators to vacuum tubes, Marshall combined pragmatism with a vivid sense of technological possibility. By the 1930s, while the company was making a variety of electrical components such as transformers and vacuum tubes (its Eveready tube became its best-known product), Marshall had also ordered research into simple magnetrons and transmitters; he even dispatched his engineers to poke into television research.

Still, by 1940, Raytheon, with its fourteen hundred employees and $3 million in sales, was insignificant compared with giants such as General Electric, Westinghouse, and RCA. When the Radiation Laboratory formed a committee to oversee its microwave radar work, it ignored Raytheon, despite the close connections the company had with working groups at MIT and the lab. Instead, representatives of Bell Laboratories, Sperry, RCA, General Electric, and Westinghouse, plus the ubiquitous Ernest O. Lawrence, manned the committee. Marshall immediately lobbied Bush, who by then had left the Raytheon board and was running the NRDC. When the British unveiled the cavity magnetron, Marshall desperately sought to be included in the manufacturing effort, or at least to be considered for the job.

In 1941, British scientist J. D. Cockcroft brought the magnetron to Raytheon's Waltham Laboratory to show it to the company's most inventive technician, Al Spencer's younger brother, Percy. Modest and understated, Percy Spencer had never attended a university, though he had already contributed to the development of the proximity fuze, a miniature tube that would trigger an explosive shell as it approached its target (typically, he had built the tiny tube for a model airplane in his spare time). Spencer questioned Cockcroft closely and examined the magnetron—he convinced Cockcroft to allow him to take it home with him for the weekend—before declaring that he thought Raytheon could not only mass produce it efficiently but improve its performance significantly (that may well be because he had seen one before: Luis Alvarez claims that he and Rad Lab colleague Ed McMillan had shown Spencer a magnetron and been invited to Thanksgiving dinner at the Marshalls as a reward. "Our visit [to Spencer] proved to be the turning point in Raytheon's fortunes," concluded Alvarez).[6] With two machinists and an assistant, Spencer soon made several basic improvements on the device and cobbled together a production line with mechanical odds and ends that included bicycle chains ordered from the Sears, Roebuck catalogue. By 1942 Raytheon was producing not only thousands of magnetrons a day but assembling full radar systems for the U.S. Bureau of Ships.

As the war progressed, Raytheon expanded furiously, driven by government funds pouring in from a number of sources. Keeping track of that money was an accounting nightmare—and a problem that Marshall would never master. By 1943 Marshall's payroll had grown four times, although the ranks of senior management had not. Field engineers were

dispatched to war zones; new facilities were quickly acquired and renovated. By 1944, employment had soared to twelve thousand. By May of
that year, sales hit $100 million, with $2.6 million in profits, and
Raytheon was running neck and neck with Western Electric, the manufacturing arm of AT&T, for leadership in the tube business. In the next
year, when the board of directors began to discuss the postwar future,
the company was churning out two thousand magnetrons a day, generating $150 million in annual sales.

Raytheon's wartime successes fueled Marshall's innate confidence.
Marshall believed the flow of military work would continue, but he, like
RCA's Sarnoff, also foresaw a vast revolution in civilian communications
that Raytheon could exploit. Marshall emphasized the scientific links
spanning technological applications and downplayed the more mundane
commercial differences. He viewed microwave communications as the
hub of a variety of new products: television, FM radio broadcasting, facsimile reproduction, nationwide police and weather-forecasting networks.
In 1945 Raytheon directors approved a change in the corporate charter
allowing the company to build high-frequency broadcasting stations and
to file an application with the FCC to begin broadcasting. Marshall and
his top managers also discussed developing cooling systems for air condi-
tioners (Charles Smith's refrigerator had not been forgotten), mobile
radios, portable recorders, and a number of other electronic products.

By early 1945 Raytheon was producing 80 percent of all magnetrons—as well as large quantities of other tubes. Its stock, listed on the
New York Curb Exchange (renamed the American Stock Exchange in
1953), had appreciated 180 times during the war, from 50 cents in 1940
to $90 a share in January 1945, although it had not paid a dividend since
1929. In February the board approved a three-for-one split; the company had sixteen thousand employees. In April, Raytheon announced
the purchase of Belmont Radio Company of Chicago, a maker of radios
and jukeboxes founded in 1930 and struggling to adjust to the incipient
age of television. Marshall sketched out plans to set up broadcasting stations across America, linked by microwave relay stations into a network
called Sky Top. As Raytheon's official history comments, "This program,
combined with Belmont, envisioned a net that was almost unworldly.
Broadcasting on one end, marketing receiving instruments on the
other—and manufacturing all the components in between—would have
left no communications sector uncovered."[7] A Raytheon advertising
campaign in late 1945 featured a large ear surrounded by the whirling
parabolic orbits of electrons, alluding to that day's favorite scientific
cliché—the atom with its cloud of electrons. "Raytheon engineers," said

the text, "expect car-to-radio telephones to become as familiar as plane-to-ground communications. . . . This is communications, [a] growing branch of the new science of electronics, which puts the electron to work through the vacuum tube and its associated equipment."

## The Pragmatism of Du Pont

If Raytheon was bold, Du Pont, the country's oldest major company and its largest maker of chemical products, was cautious. If Marshall saw Raytheon crashing through barriers, top management at Du Pont felt for doorways. If Raytheon saw the commercial possibilities in technology, Du Pont sensed pitfalls and problems.[8]

Du Pont feared the coming of the war. Not only would wartime shortages hinder the explosive growth of nylon, but experience had taught company executives that, although wartime work could be financially profitable in the short term, it could be politically dangerous in the long term. In the mid-1930s, the congressional Nye Committee had labeled Du Pont executives "the merchants of death" for the profits the company reaped as a World War I munitions supplier. Liberal sentiments against giant corporations such as Du Pont still ran strong, and Du Pont was often attacked for its participation in prewar international cartels with German firms such as I.G. Farben. As we have already seen, the upsurge in antitrust activity forced Du Pont executives to abandon acquisition as a means of future growth and seek out research as a way of developing new products. As early as 1941, with Pearl Harbor still months away, Du Pont president Walter Carpenter began to fret about the company's *postwar* readjustment. Angus Echols, a member of Du Pont's executive committee, laid out a detailed strategic plan that focused on using an expanded research program to generate new, high-profit products from organic chemistry (the allure of nylon had already captivated Du Pont executives), thus allowing the company to get out of the arms and munitions business. Echols recognized how the world of business and research had fundamentally changed since 1939. "The wartime conditions made all this more urgent. Echols saw the federal government 'going into business in a big way' and the wartime emergency pushing many other companies into new areas. He urged that Du Pont plan its postwar future immediately."[9]

Unlike Raytheon, which took any business it could find, Du Pont attempted to discriminate; it refused to remake itself for the war. The Manhattan Project posed a particular challenge. Early on, General

Groves sought out Du Pont to take pressure off the overburdened Stone & Webster. Groves asked Du Pont to build the pilot atomic reactors needed to produce bomb-grade plutonium—an alternative track to the Oak Ridge Y-12 works. The project quickly mushroomed into the Hanford Engineering Works, a massive plutonium facility built along a barren stretch of the Columbia River in eastern Washington State. Du Pont presided over Hanford as a sole contractor, fitting it into its already complex organizational structure as its TNX division; eventually Hanford would employ forty thousand souls and pose monumental management challenges. Clifford Greenewalt, Du Pont's head of R&D, had to act as a liaison between Arthur Compton's Metallurgical Laboratory in Chicago (the site of the first atomic pile and an R&D center like Los Alamos and the Rad Lab), General Groves's military staff, and Du Pont senior management, while also supervising the rapid construction of the working complex itself. Greenewalt's performance at Hanford was an engineering triumph; Hanford, unlike Y-12, would continue to churn out plutonium for decades. By 1943 Du Pont had not only a major new division but an unparalleled opportunity to dominate an exciting new postwar business. By the end of the war, no other company on earth had the detailed knowledge of state-of-the-art atomic power that Du Pont had.

And yet, as early as 1942, Du Pont executives began to debate whether they should pursue nuclear power and engineering as ongoing business lines after the war. Company executives quickly concluded that the only real business of atomic power would be in power generation—a business Du Pont had never participated in and one, with its heavy government regulation, that the company would normally resist. There was also among the senior executives at Du Pont a recognition of two limitations: first, that the work in this area still was quite rudimentary, suggesting the need for long lead times and massive expenditures on R&D, and second, that the government would be proprietary when it came to atomic technology. Would the government allow Du Pont access to patents, know-how, classified data developed at Los Alamos, Hanford, or Oak Ridge? Would the government wish to keep much of this knowledge secret, wreathing any business in security and red tape? Would Du Pont have to be in constant consultation with government officials and face close oversight by potentially hostile congressional committees? In short, what kind of company did Du Pont want to be: one battling in consumer markets or one attached by a short leash to the government?

By 1944 Greenewalt, an MIT-trained chemical engineer deeply involved in the nylon developmental project who would later serve as Du Pont's president, had developed more detailed arguments for abandoning nuclear power. He had spent the war years wrestling with the intelligent, mercurial, egocentric Chicago physicists—brilliant, occasionally maddening personalities such as Szilard, Fermi, Eugene Wigner— and he knew how opposed most of them were to corporate involvement. They were a different breed from the chemists who, over the years, had grown accustomed to working in a corporate harness. Building a postwar research effort with them would be difficult and expensive; trying to do it without them might be fruitless. Moreover, no applications beyond power generation had yet appeared; thus, the rate of return on its investment (and Du Pont's sophisticated senior executives kept such figures close at hand) appeared minimal for some time. Even if patents could be obtained, R&D would be so protracted that the company might not be able to take advantage of them. Greenewalt's conclusion: "Du Pont was unlikely to ever recover the money it spent on nuclear physics."[10]

Although there was considerable debate within the company, Du Pont's general assessment of atomic power ran counter to the glorious predictions for peaceful applications of such power made after the war. In May 1946 Du Pont told Groves that it wanted to withdraw from the Manhattan Project contract in order to return to a focus on consumer products. Du Pont recommended that General Electric, already a major supplier of electrical power generation equipment, take over Hanford. (General Electric did so, becoming a major vendor of power plants in the decades ahead.) Several times over the next few years, the government approached Du Pont to take on national security projects, particularly the building of the breeder reactor at Savannah River, Georgia. Du Pont always acceded grudgingly, apparently with few commercial ambitions in that area.

The differences between Raytheon and Du Pont would be a common motif in the postwar years. Raytheon's Marshall saw different business lines as fundamentally similar; an underlying technology, in this case electronics, dissolved differences in those businesses—military versus civilian, service versus industrial, wholesale versus retail. Developmental cycles, funding requirements, and the mind-sets required of managers meant less to Marshall than that each business employed vacuum tubes and wires and "put the electron to work." In later years, proponents of this point of view would argue how businesses as different as television and radio broadcasting, consumer electronics manufacturing, radar, communications systems, and missiles and microwave ovens (a Raytheon invention) were synergistically linked. Synergy would become the rallying cry for technological bulls, not only in entrepreneurial firms but at the corporate conglomerates of the 1960s and,

eventually, in a particularly pure form undiluted by scientific knowledge, on the Wall Street of the 1970s and 1980s.

Du Pont, bearing its burden of age, pedigree, and size, had a much
different perspective. The company was excited by scientific possibilities and recognized the necessity to invest in them; in fact, it would bet
the company's growth in the postwar years on costly scientific research.
But where Raytheon and RCA embraced technological homogeneity,
Du Pont drew away from what it perceived as deep differences.
Nuclear power promised profits only in one business, power generation, which Du Pont knew nothing about and which would never fit
easily into its current, chemically oriented lines. Certainly, nuclear
power was a burgeoning field, but how quickly would the profits come?
After several decades wrestling with the complexities of organic chemistry, Du Pont executives were not sanguine about tackling a novel,
immature technology only recently (and perhaps prematurely) yanked
from its scientific womb. And as a result of its painful past experiences
and its own strategic calculations, the company wanted nothing to do
with markets dominated by the government.

Did the split between Raytheon and Du Pont represent a division of
experience between a relatively new (and smaller) firm and a much
older (and larger) one? Was Du Pont blinded by bureaucratism and
Raytheon freed by Marshall's entrepreneurial vision? Or was it a split
between a company dominated by electronics and one based upon
chemistry? In fact, all those hypotheses reduce a complex situation to
caricatures. Du Pont was large and hierarchical, but it had proven capable of reacting nimbly to new research in chemistry; and it may well
have been the most innovative business organization in America
throughout the previous thirty years. Moreover, it was not as if Du Pont
had wandered much further afield than Raytheon. Atomic physics had
as much in common with Du Pont's traditional chemistry R&D as, say,
broadcasting had in common with Raytheon's vacuum tubes. And it
wasn't as if Raytheon was all that more technologically focused than Du
Pont. Although Raytheon was much younger, both firms owed their
rapid growth rates in the 1920s and 1930s to emerging technologies.
Chemistry, it is true, had begun to mature more quickly than electronics.
With hindsight, one can see that the golden age at Du Pont came in the
1930s, when the company first commercialized its breakthrough synthetic fibers. For all their forecasts and calculations, Du Pont executives
had no way of knowing that organic chemistry would require more and
more capital to produce incremental growth in the postwar years. And
although Du Pont stumbled in the decades to come looking for "a new

nylon," management did prove to be quite prescient about the dangers, and limitations, of nuclear power generation.

A more fundamental gap separated the two companies. Although Du Pont was still family controlled in 1945 (Greenewalt was married to a du Pont, Carpenter was a du Pont cousin), the company was obsessively concerned over its relationship with an outside world that included not only shareholders but the government. A new, powerful, and uncertain technology such as atomic power posed a threat to those outside relationships. Shareholders, including many du Pont family members whose current and future fortunes were tied up in the stock, faced taking smaller returns in exchange for a future windfall—an uncertain future windfall—if the atomic option were chosen. More important, Du Pont's very domination of such a business would inevitably bring the government down upon it. Du Pont had much to lose from continued controversy, particularly its lucrative stake in General Motors.[11] And there was the competitive threat. Du Pont executives recognized how the war had diffused proprietary research and exposed smaller competitors to large-scale projects. Those companies—such as Monsanto, Union Carbide, Allied, and Dow Chemical—would be far greater competitors than in the past, and Du Pont decided it did not need the distraction of a new business. Finally, organic chemistry simply seemed to be able to provide all the growth Du Pont could handle.

Raytheon's Marshall operated in a much simpler psychological environment. Despite his extensive ties to the military, Marshall lacked the intense concern with how outsiders viewed his company; in that, he displayed a classic entrepreneurial attitude. Raytheon stock had not been valuable long enough for shareholders to develop a sense of the company as a legacy, although the Morgan interests kept a seat on the board. Moreover, the very presence of the government served to balance the demands of the public markets. With government orders pouring in, Marshall did not seem to think he needed the public debt or equity markets as a primary source of financing; by 1945 the company was essentially funding itself. That being the case, Marshall could ignore Wall Street complaints that Raytheon paid tiny dividends, always seemed to be restating its earnings (its books were a disaster until a young financial genius named Harold Geneen arrived to overhaul them in 1956), and was constantly trying to enter new businesses. As for the competition, Marshall's confidence had shaded into hubris. If Raytheon could humble mighty Western Electric, what feats couldn't it achieve?

By 1950 Marshall's vision was in trouble.[12] Technological skills could not overcome funding shortfalls, and the company began to lose the footholds it had established in civilian businesses such as television broadcasting, microwave communications, and radio. Belmont, in particular, became a cash drain as the television-set wars commenced, and even its military work became more precarious as development on missile and radar programs was speeded up. Marshall's response to these difficulties was to find a merger partner that could provide the funds to pay for all of this. In 1949 he approached ITT, the global telephone company founded and run by the flamboyant Sosthenes Behn. In January 1950 Marshall approached the Raytheon board of directors with the plan, only to encounter deep skepticism. The board handed the project off to the company's director and president, Charles Francis Adams, Jr., the patrician and low-keyed scion of the Boston Adamses and a man with a more intimate understanding of the concept of legacy than Marshall. Adams had been installed on the Raytheon board in 1938, at age twenty-eight, to watch over the Morgan interests (Henry S. Morgan had married Adams's sister). In 1947 the board named him executive vice president; a year later he was president and CEO. ITT eventually dropped the merger plan, and Adams quietly assumed power, beginning the process of rebuilding Raytheon. In May 1950, Marshall retired, immediately leaving Boston to lead an expedition, on behalf of Harvard's Peabody Museum, to study the life of the Kalahari Bushmen.

## Kilgore versus Bush

Another manifestation of the split between optimists and pragmatists unfolded in wartime Washington, among policymakers struggling to forge a role for the government in this new scientific age. By 1943 Congress had actively begun to wonder what the postwar years would be like. Characteristically, questions arose from current complaints and past concerns. Considerable criticism had surfaced over the small number of large corporations that seemed to be receiving the bulk of the government's military contracts—a radically different perspective from Du Pont's view that the war would sharpen competition. A Senate subcommittee on small business produced statistics that showed that the government paid $175 billion from 1940 through September 1944 to 18,539 companies in wartime contracts. Two-thirds of these contracts, however, went to the one hundred largest U.S. corporations and 30 percent went to the top ten companies.[13] In Washington, the successes of relatively small firms such as Raytheon were rare indeed.

Such statistics raised the familiar specters of monopoly power at home and cartel power abroad. "Concentration of economic power is still a menace, and may well be a growing one," said the *American Economic Review* in a review of a book about small business. "While winning the war has greatly relaxed our alertness, it has not removed the danger. The very industrial gigantism fostered by huge war contracts, however necessary, may, unless neutralized, nourish speedily within our borders the same enemy 'new order' of monopolistic, cartelized statism which cost so much blood and treasure to exterminate."[14] In the liberal *Nation,* I. F. Stone began a prolonged campaign against "the cartel cancer" in 1944. He drew out all the threats posed by domination of a few large corporations: the return of an economic debacle such as the Great Depression through the predations of domestic monopolies and the revival of the Nazi menace by allowing international cartels to survive. Technological might was particularly at risk. "Cartels," declared Stone, "either stifle technological development or hog its benefits." In an article several weeks later, Stone advocated that the twenty-six hundred federally funded war plants be retained by the government and used as a competitive counterbalance to the large corporations.[15]

Stone's subject in his article "The Cartel Cancer" was the work of a Senate subcommittee on war mobilization presided over by a round-faced West Virginian and New Dealer named Harley Kilgore, an attorney, judge, and World War I veteran. Soon after Kilgore was elected in 1941, he was asked by an old friend, Sen. Harry Truman from Missouri, to join his War Investigations Committee; Kilgore in turn formed a subcommittee to study war mobilization issues. Kilgore then transformed that subcommittee into a bully pulpit, investigating how the government was running the war, holding hearings on war contracts, and drifting to longer-range issues dealing with postwar policies, from cartels to scientific and technological policies. "American liberals have a larger debt of gratitude to Senator Harley M. Kilgore (D., W. Va.) than many of them realize," editorialized the *New Republic* in January 1945. "He has worked hard on the problems of international cartels, domestic monopolies in railroads and communications, and postwar economic collaboration. We have him to thank, in part, that the official plans for reconversion, certainly not very good, are not a good deal worse."[16]

In late 1943, Kilgore had begun to formulate plans for postwar science and technology policy.[17] Like Stone, Kilgore wanted the federal government to be involved actively as a counterweight to private, corporate, potentially monopolistic big business interests. Kilgore was impressed by the mobilization of scientists during the war, and he sought to extend that mobilization into peacetime. His first Science Mobilization Bill, S. 702, proposed a National Science Foundation that would, like Bush's OSRD, control and manage resources pouring into science and technology (Kilgore chose the term *foundation* to undercut the prevalent fear among scientists of a government bureaucracy). The president would choose the head of the NSF, who would in turn appoint a board of advisers. Patents generated through government's direct research or its funding would fall into the public domain. Kilgore's bill also urged that a percentage of the effort be poured into social sciences, and he advocated not only seats on his board for small business and consumers but direct aid to entrepreneurs. Some of the money would also be disbursed geographically.

Kilgore's bill quickly attracted comment. Industrial representatives claimed that an NSF would threaten its own research efforts and undermine the patent system. Academic scientists feared that the NSF would politicize science, create a new and cumbersome bureaucracy to deal with, and siphon off money to "soft" social science projects. Biomedical researchers feared physics would get most of the money, physicists feared getting drafted into more weapons research, and the army and navy feared they would lose power to a civilian agency. Many of these critics found an ally in Bush, then running the OSRD and, except for the president, the most powerful science and technology policymaker in the government.[18] Bush believed that the government should continue funding R&D after the war, but he disagreed profoundly with the political framework Kilgore sought to erect to do that. Kilgore, however, appeared eager to compromise. He deleted parts of the bill that scientists found coercive and invited Bush and his OSRD staff to help hammer out a final draft.

In Bush, Kilgore came across an opponent who was a sort of scientific jack-of-all-trades: teacher and administrator, bureaucrat and entrepreneur, politician and technocrat, scientist and engineer. On the one hand, he could pose like some figure out of a cracker-barrel past, particularly before a congressional committee—a crusty old Yankee who had grown up a clergyman's son in tough Irish Back Bay Boston and who preached individual initiative and enterprise. On the other, he operated in the daily flux of the scientific present: an inventor, the developer at MIT of the differential analyzer (an early analogue computer), a man whose entire career as teacher, consultant, bureaucrat, and administrator rested on his talent for putting effort and money into scientific initiatives with the greatest technological potential. "He seemed to flourish in the secrecy and tight compartmentalization of science imposed by the military, and as time passed by his grey hair grew more professionally awry, his Yankeeism and air of mystery more settled," wrote *Fortune* in 1945. "He had his official picture taken in a crisp laboratory smock, pipe in mouth, holding up an artfully lighted measuring glass of clear liquid. This slightly Hollywoodish image of a scientist was useful on Capitol Hill."[19]

Above all else, Bush was an engineer—a very sophisticated one with a strongly mathematical bent, but one who also placed himself directly in the tradition of Edisonian invention. Such an attitude allowed him to appreciate and foster not only the entrepreneurial efforts of Marshall but the organizational labors of Groves and his Manhattan Project. "Much of what OSRD accomplished was engineering, not science," writes science historian Nathan Reingold, ". . . but engineering with rather distinctive attributes by the standards of the day."[20] His academic field had been electronics, particularly the new realm of computing machines, an interest that operated on the frontier between the abstractions of quantum mechanics on one side and the tangibility of electrical engineering on the other. Thus, like Sarnoff and Marshall, though without their fever, Bush was sensitive to the potential in electronics. He had also spent decades at MIT, with its mixture of science and engineering and its complex ties between academia and industry. In fact, Bush was a younger colleague of Dugald Jackson, who established the MIT electrical engineering curriculum (Bush later said that when Jackson came to speak to his introductory electrical engineering class, the students complained that they were "learning a great deal about public utility companies and their management," but little about electronics).[21] And of course his role at Carnegie and in the government gave him wider perspectives than the run-of-the-mill academic administrator. Bush believed strongly in the necessity of basic science—and of the economic necessity of providing its technological fruits to industry.

Bush quickly set out to learn Washington politics, where he juggled an appearance of stubborn independence with real talent and enthusiasm for consulting the opinions of others. He worked directly for Franklin D. Roosevelt, operated within the frame of New Deal politics, but never liked nor trusted New Dealers (though he did admit to a loyalty and affection for FDR and admired his liaison, Harry Hopkins). Ironically, Bush was personally more sympathetic to the worldview of Herbert Hoover, the Great Engineer, with his notions of private enterprise and a less interventionist government. Although Bush saw the need for the expansion of government power during the war and was more than willing to serve as its agent, he also attempted to act in a way that left as many traditional power structures and institutions standing as possible. For the most part, Bush succeeded in making OSRD into a flexible, effective bureaucracy, and he did it not with pronouncements from above but through a constellation of committees and panels very similar to the densely woven network of faculty, foundation, and engineering associations he had long dealt with. (He wasn't perfect, though: he rebuffed Robert Goddard during the war and denied for a long time the feasibility of rocketry.) And as the war came to an end, Bush attempted to do what so rarely ever happens in Washington: dismantle the bureaucracy that gave him such power.

By mid-1944 Bush, too, was thinking about how postwar science and technology might be organized, particularly about how he could eliminate what he saw as the expedient, if dangerous, vehicle of the OSRD. The administration was gearing up for an election and looking for a way to publicize its achievements. Bush's OSRD was an obvious candidate. Hopkins and Oscar Cox, an attorney and Treasury official who had first introduced Bush to Hopkins in 1940 and later became Bush's personal lawyer, came up with an idea for Roosevelt to ask Bush to write a report describing his ideas on postwar science and technology.[22] Bush eagerly agreed, then began a series of meetings with Kilgore, leaving the impression that his proposal would represent a compromise between the two points of view. Instead, on July 19, 1945, with Bush out of town, the OSRD released its report and plan in a book called *Science, the Endless Frontier*. Washington's freshman senator, Warren Magnuson, introduced the bill in the Senate and, at the same time, young Arkansas congressman Wilbur Mills presented the bill to the House. Kilgore felt double-crossed.

Bush's plan received enormous publicity. *Fortune* condensed it into a special supplement, and the book itself became a best-seller. Like Kilgore, Bush sought an NSF funded by federal money. But unlike the senator, he sought to leave control in the hands of private interests—notably, the scientific and engineering communities. "I have come to the realization," he wrote in a letter to *Science* magazine in 1944, "that science flourishes to the greatest degree when it is most free."[23] To Bush, centralized control was necessary in war, damaging to science and to American freedoms in peace. He proposed that the president appoint a board of experts without government ties, which would then appoint a director. There would be no restrictions on patent policy, which would be controlled by the board. Bush, perhaps recalling the early days of Raytheon, strongly believed that innovation occurs most often at smaller units and that those units must have patent protection (he did not, however, propose funding for small businesses and entrepreneurs as Kilgore did). Social science, which Bush had little respect for, was ignored, overwhelmed by an emphasis on supporting basic research, mostly through support of existing institutions. And although the report had emphasized support for medicine, it had said very little about the biological sciences.

Thus commenced a long debate that would not be resolved until the early 1950s. Although Bush, a giant of wartime science, had become famous—and, through *Science, the Endless Frontier,* virtually synonymous with science policy—Kilgore and his allies in the administration held the political cards. When Truman succeeded Roosevelt in 1945, Kilgore had a friend in the White House who was not averse to keeping the reins of power in his hands; Bush quickly lost access to Truman. Moreover, Kilgore skillfully orchestrated hearings that rang still-potent New Deal chords. "I get a little tired of these hired hands of the monopolists and some of the professors, some of these bulldozing scientists, piously abrogating [sic] to themselves all the patriotism," testified Maury Maverick, chairman of the Smaller War Plants Corporations. "Their superior attitudes at least become obnoxious to me if not to other people." Maverick added, "The moral character of politicians is just as high as the moral character of the American scientist."[24] Kilgore himself weighed in on the patent issues. "It seems to me when taxpayers of the United States pay for the development of something, it is a crying shame to make them dig deep into their pockets and pay a big royalty to some outfit that has grabbed the results of their research."[25] When Truman finally signed the NSF legislation in 1950 (after vetoing an earlier bill in 1947 over the issue of presidential control), many aspects of *Science, the Endless Frontier* had been watered down or eliminated. The president retained the right to appoint the NSF director, though he would share power with a board, not dominate it; patent policy remained much as it had during the war; the legislation mandated money for social sciences; and geographic diffusion had been dropped. The new NSF was set up mainly to pursue pure research and was not associated with the military.

Thus Bush failed to get much of what he had sought. In a suggestive essay on Bush written in 1987, Reingold argues that Bush harkened back to an age (and a set of ideas) that had become, like Hoover, irrelevant.

> His was a striving to hold back a future with blurred boundaries between public and private, between contemplation and action, between power and ideas, between national goals and self-aggrandizement. His was a worldview of discrete "local" entities, largely self-governing and self-motivated, and each requiring careful attention to their intrinsic natures. It was a conception of minimal government and of individual initiative—of personal responsibilities and attention to others. Striving to work within the system, Bush often endangered the consistency of purpose as he tried to compromise with others.[26]

But were Bush's ideas really irrelevant? Had they really died? The fact is that, as Kevles points out in his history of American physics, the NSF never became the dominant funding mechanism that Kilgore proposed. The landscape that evolved was far more complex, crowded with private and public institutions—some closer to the Kilgore model, like the NSF, others leaning toward Bush's pragmatic approach, all jockeying to support science and technology. Neither man foresaw the full effect of the money they were about to unleash. That money spawned large, noisy scientific and bureaucratic interest groups that were quite capable of making their demands heard and influencing research policy. Moreover, the very complexity of the science put politicians at a disadvantage. Most decisions involved recondite scientific issues well beyond what even interested politicians, or their staffs, could handle. Only a review system dominated by scientific peers—administered by specialized bureaucracies—could keep the machine turning. Even when political dicta did come from above, such as President Nixon's 1972 decision to pursue the war on cancer (an initiative Bush privately opposed), biologists were able to seize control of the fund-raising mechanism, in this instance, at the National Cancer Institute. In fact, the war on cancer boosted the rise of molecular biology, which in turn spun off the biotechnology industry with its hordes of small companies.

## New Alliances

All of these personalities, of course, believed deeply, and for the most part uncritically, in the merits of continued technological progress. Their differences, which are considerable, clustered around questions of degree and control; and so they eventually became differences of politics as well as of policy.

Despite the ties of friendship and business joining Bush and Marshall, it was Marshall and Kilgore who shared the tremendous optimism about science and technology's ability to reshape the postwar world—and to reshape itself. They saw science as a kind of Platonic substance running beneath a world of appearance and delusion. One need only understand the nature of that scientific substance to tap, channel, and exploit its power; and once released, it would sweep superficial limitations and barriers aside: a true World of Tomorrow. That understanding, they believed, had been achieved. Both Kilgore in government and Marshall in business viewed science and technology as the means to reorder, renovate, and reengineer the world for public betterment or for private profit—or, as Adam Smith suggested, for both. One telltale sign of this attitude at work is the continuing emphasis on delineating new epochs and ages, each of which suggests a deep break with the past. The knowledge now at hand allows us to create a new epoch, with new human material.

On the other hand, to read Bush is to wade through the kind of distinctions Charles Stine generated to convince Du Pont's executive committee on the merits of pure research. Applied and basic science, academic science, industrial R&D, government research—all had complementary, if very different, roles to fill. Distinctions and differences mattered to Bush. Technology was not a well to be tapped but a marble block to be chipped away at, polished, struggled against. Science was a kind of art—and a difficult one at that. Its creation could not be ordered, like dinner in a restaurant. Despite his sympathy for entrepreneurial energy and small innovative units, Bush instinctively distrusted talk of transformations; in the years ahead he would warn about "crass misconceptions of science" by people who believed that scientists were supermen who "can do anything, given enough money."[27] Bush believed science needed to be nurtured, that technology could create great good. But he neither desired nor quite believed in technological revolutions or a golden age of science.

It is one of the defining traits of the postwar era that none of these different ideas would be washed away in the decades ahead; they would promiscuously cross-fertilize and form new combinations. Kilgore's vision of big government and big science would produce the space program, the war on cancer, and Star Wars—and the demands both for science mobilized against social problems and of an industrial policy. Bush's appreciation of local entities and small innovative units would return in the guise of free-market economics and, more tangibly, with the rise of the venture capital business. Marshall's view of the technological imperative would eventually feed the messianic strain that appeared in American commercial technology and leave corporations such as Du Pont, with their structures and studies and discriminations, feeling increasingly besieged and outmoded.

# 4

# The Conservatism of
# Television

## The Ambiguity of Television

In 1946 television came of age. The then-popular books on the subject capture the giddy sense and wonder that was part and parcel of the postwar fascination with science: *Television, the Eye of Tomorrow* (1942), *Television, the Revolution* (1944), *New Television, the Magic Screen* (1948), *The Miracle of Television* (1949), *Television, the Magic Window* (1952). Public officials also caught the fever. "A new force is unloosed in the land. I believe it is an irresistible force," declared FCC head Wayne Coy in 1948.[1] "Television is the most dynamic element in the entire American economy," *Fortune* added in 1949. "Within a few years—perhaps five—it will be one of the first ten industries in the U.S.; already it is the great American Adventure of the mid-Twentieth Century."[2]

In the first two years of television, 1946 to 1948, Americans spent between $750 million and $1 billion for sets, a liftoff that RCA president Frank Folsom declared the fastest in American business history.[3] Television, he added, reached the billion-dollar level five times faster than the auto industry had earlier in the century. What made this consumer explosion even more remarkable was that while some five hundred U.S. companies manufactured televisions in the first great selling wave, there were still only sixty-six television stations across America, and the broadcasting industry was still struggling to provide coast-to-coast, daily programming.

Despite the excitement, mounting sales, and intense competition, television would prove over time to be less than revolutionary in terms of its structural effect on consumer electronics. Although television momentarily stunned radio ("Radio is the only American industry to die in its infancy," predicted Milton Biow, an advertising executive),[4] it did not, in fact, kill it, and it did not fundamentally transform the communications landscape. AM and FM radio would each find a profitable niche in an expanded spectrum of electronic entertainment next to television, and, after a brief dip in the early 1950s, radio-set sales would again grow. Television was undoubtedly a major triumph, and it dominated mass consumer electronic entertainment for decades. But if you define technological revolution as killing off the previous regime, television was not revolutionary at all.

Despite its novelty, the American television industry looked back far more than it peered forward. The television industry that sprang up after the war more strongly reflected the companies at the World of Tomorrow than, say, the transistor companies that were appearing at the same time. Technologically, television was dominated by the vacuum tube and was thus an extension of traditional radio technology (the bulbous television tube was actually a blown-up vacuum tube and, indeed, injected new, if brief, life into the tube business in the early 1950s).[5] Historically, television had been dreamed up around the same time as radio, though its progress in consumer markets was much slower. The industry was limited by a number of factors. Financially, the television industry was formed in an era during which there was very little external capital available for new ventures; the advantage went to large manufacturers with internal capital resources. Television was also limited to a fairly narrow evolutionary track by regulation, not only in terms of antitrust but, just as important, in terms of mandated FCC technical standards. Thus, despite the mob of early companies, mounting capital costs (for both R&D and marketing) and limited venture capital dampened future entrepreneurial fervor. The industry quickly formed an oligopoly made up of large manufacturers; and, relatively speaking, it was quite stable.

The television industry thus provides a fascinating benchmark upon which to contrast more volatile, faster-changing technologies born in a more expansionary financial age. The central truth about television is that despite its sophisticated electronic heart, it was a conservative technology from birth. And the telltale signs, the phenomenology, of that conservatism—not to mention the symptoms of the demise of the American industry—lay in the oligopolistic industrial structure that quickly arose. Structure, in this case, directly mirrored technological evolution, not revolution.

## A Technology Delayed

The first device for receiving signals from a transmitter, which established the feasibility of television, was patented in 1907, the same year Lee de Forest invented his three-element vacuum tube, the audion, which made commercial radio possible. While television struggled to untangle the technological kinks, radio took off. Although a handful of mechanical television technologies were concocted in the 1920s, notably at Bell Laboratories, the commercial breakthrough came in 1923 when a Russian immigrant from St. Petersburg named Vladimir Zworykin developed a rudimentary television camera, the iconoscope, followed six years later by a receiving tube, the kinescope. Zworykin went to work at Westinghouse, which only half-heartedly supported his efforts. When Sarnoff, himself born in czarist Russia, orchestrated the independence of RCA in 1932, he grabbed Zworykin and gave him a lab, some money, and access to RCA's busy patent attorneys.

While Zworykin was inventing his iconoscope, a Mormon inventor from Utah with the Capraesque name of Philo Farnsworth came up with a similar invention, which he called the image dissector. Farnsworth went to California, interested a few local bankers—this was still the 1920s, when shares of electronics firms such as RCA were driven to dizzying heights on Wall Street—and set up a company to develop a patent position. The first image he successfully transmitted was a dollar sign. In the early 1930s, Farnsworth was momentarily saved when Philco, the radio manufacturing firm, backed him with $250,000 over several years to perform television research, before deciding that a commercial product was still some way off and cutting him loose. Farnsworth struggled along on his own again in Philadelphia before his shareholders forced him to set up a manufacturing arm in 1938, Farnsworth Radio & Television, to make radios and defend his now impressive collection of patents against litigation, mostly from RCA. Farnsworth won most of the cases, but, never the enthusiastic businessman, he decamped for a laboratory in Maine in the early 1950s and died, loaded down with 350 patents, in 1971.

Two markets (radio and television), a handful of complex technologies, dozens of companies, the government, patents, the lure of windfall profits—all provided the ingredients for a long, spicy struggle over television regulation. It is a mistake to assume, from a 1990s perspective, that this struggle was one between corporations seeking to avoid regulation and bureaucrats eager to shackle commerce. No one in either government or industry wanted to kill the goose that lay the golden egg of television. Businessmen envisioned huge profits, political liberals saw a further expansion of democracy and uplift, and regulators sought service that would expand in a rational, orderly way and at the least cost to consumers. At the very minimum, the microwave spectrum was a limited public resource that had to be parceled out in some rational way. Regulators thought they were protecting consumers, broadcasters, and manufacturers by eliminating incompatible systems; despite the seemingly endless wrangling, all parties generally agreed on that. Most also thought it good public policy to force RCA, that potential monopolist, to license its patents out at reasonable prices. Even RCA chairman Sarnoff, eager to diffuse the technology as quickly as possible to reap a harvest of royalties and build an audience for NBC, RCA's broadcasting arm, eventually agreed and moved to facilitate licensing.

The regulatory game consisted of one major player, RCA, and a host of secondary players, including broadcasters such as CBS, DuMont, and Mutual and manufacturers such as Zenith, Philco, Farnsworth, and the rest. The commercial goals were to preserve radio sales (or broadcast revenues) and, particularly for Sarnoff, to keep those radio royalties coming, until it was the "right" time for television. Sarnoff, with investments in both broadcast and manufacturing, had a particularly tricky task. No one would watch television, including shows on NBC, if television sets were either crude or expensive; no one would buy sets if there was nothing to watch. However, Sarnoff also wanted to maximize the value of his patent position, without allowing others to chip away at it. For most nonbroadcasting radiomakers, the optimal time for television was later—or never; their strategy was to lobby hard for higher-quality standards than Sarnoff wanted. In 1938 Zenith declared in its annual report: "The management continues to believe that television is not ready for the public and refuses to be stampeded into the premature production of television receivers for sale to the public."[6] Philco claimed a victory when the FCC chose 525 scanning lines per second, well above the 441 RCA was urging but below its own proposal of 604. As for the broadcasters, CBS and DuMont were not leaping to license RCA technology, buy RCA equipment, and help RCA pay for both future color research and big stars for NBC.

Even as the FCC struggled to come to a decision over black-and-white television standards, CBS lobbed into the debate the issue of color. CBS under William Paley had acquired a mechanical, wheel-based color television system developed by a Hungarian engineer named Peter Goldmark. Paley had no way of manufacturing and distributing the sets, but he did wish to slow the RCA juggernaut—and color provided the means. In an attempt to dislodge RCA, CBS argued that the FCC should approve its color-wheel technology, even though it was incompatible with RCA's monochrome system (RCA had a compatible color system under development, but in the early 1940s it could not match the color quality of the CBS system, and Sarnoff did not want to cannibalize potential black-and-white sales with color). Although the CBS mechanical color system would eventually fade—in the 1950s the FCC set standards for an electronic color system—black-and-white was again delayed.

Making the situation murkier, frequency modulation, or FM, for radio also surfaced in the 1930s. Here was additional evidence of mounting technological competition. FM threatened to crowd the VHF spectrum RCA hoped to use for television, thus posing a threat not only to its radio business, built on AM patents, but to its television patent position. Sarnoff, who had once backed FM inventor and developer Edwin Armstrong, refused to act on FM, with its static-free signal, until the late 1930s (he was lobbying for AM sound for television), thus enhancing his reputation as a "patent racketeer." After protracted lawsuits, Armstrong, once a close friend of Sarnoff and a major RCA shareholder, leapt to his death from New York's River House in 1954 just after agreeing to an RCA settlement.

All these forces delayed American television for years. As early as the mid-1920s, Sarnoff and others were predicting its imminent arrival. Television demonstrations were held in department stores and radio outlets and at trade shows throughout the 1930s; newspapers were full of tales of Sarnoff, Farnsworth, and Zworykin. Nazi Germany began commercial broadcasting in 1935 with a crude mechanical system, and Britain launched a superior electronic scheme a year later (at decade's end there were over twenty thousand sets in London alone).[7] By 1939 a handful of companies were selling sets to receive experimental broadcasts. By then Sarnoff was ready to force the issue. With only two hundred sets in New York, RCA's television launch at the World of Tomorrow was less an attempt to commercialize television fully than a gambit to present the FCC with a fait accompli. In the short run, the effort flopped. The war arrived, an audience failed to develop despite marketing incentives, and the FCC under New Dealer James Lawrence Fly, a former general counsel for the Tennessee Valley Authority, charged RCA with monopolistic tendencies and forced it to suspend broadcasting. RCA then joined an industrywide panel, the National Television Systems Committee, to hammer out standards, which were then passed on to the FCC. In 1946, the FCC, under attorney Charles

Denny, approved the NTSC's black-and-white standard and commercial television broadcasting began again, this time for good. Sarnoff then brazenly hired Denny to be his general counsel.

## The Uneasy Reign of RCA

RCA was clearly the most powerful financial force in the nascent television business. It was not only the largest of the consumer electronics firms—1947 sales topped $314 million, with a net profit of $18.8 million—but it possessed a brand name synonymous with radio, it had a major broadcasting unit in the form of NBC and extensive business lines in radio communications, and it had Sarnoff. Born in the Jewish pale near Minsk, raised on the Lower East Side of New York, Sarnoff had gone to work in his teens as a wireless operator at American Marconi, where he made himself famous for playing a much disputed role in receiving the telegraphic news of the *Titanic* disaster. When RCA was first organized, the still-young Sarnoff was appointed to run what was envisioned as a conservative, defensive operation. Instead he aggressively augmented RCA's patent position, transformed the radio from a clumsy, balky device for long-range communications into a mass consumer product, and noisily began television research—his "radio with sight."

Sarnoff reinvented himself and RCA into sleek symbols of a new, electronic age. He was the very model of the technological entrepreneur, sketching out, like Raytheon's Marshall, great dreams of RCA in new products and new industries; in so doing, he made RCA appear mightier than it really was. But, unlike Marshall, he was also the ultimate pragmatist, willing to bury or block technologies, like FM, to advance RCA's interests. Sarnoff was a larger, more complex predecessor not only to Marshall, but to that icon of the 1980s computer industry, Apple Computer's cofounder Steven Jobs. Neither Jobs, Marshall, nor Sarnoff invented anything. All three were essentially men—young men during their initial success—driven by a single idea. Sarnoff and Jobs each took a complex technology (the radio and the microcomputer) and repackaged it for mass consumption. Both were proclaimed in the press as visionaries; both were quixotic, autocratic self-promoters with avid followers. While Sarnoff was far more successful than Jobs, both were identified with companies that were strong on elegant, technical solutions but paradoxically weaker on the nuts and bolts of manufacturing and retail marketing; they were entrepreneurial rather than managerial.

Certain historical RCA weaknesses made market control difficult, despite Sarnoff's sweeping rhetoric. Much of the earliest regulatory effort swirling around television was made in direct response to fears that RCA would monopolize the television as it had radio. Indeed, RCA was actually founded in 1919 as a monopoly at the urging of the U.S. Navy, which thought that a single vehicle for the radio patents of General Electric, Westinghouse, and American Marconi, later to be joined by American Telephone & Telegraph and United Fruit, would best serve national security purposes. But as political winds changed, so too did the government's attitude toward RCA's market dominance. Over the next twenty years the radio industry was wracked by fierce patent suits, most involving RCA.

In the 1920s, RCA fought suits stemming from its practice of licensing its patents at exorbitant prices and with conditions that seemingly ensured its preeminence (the right, for example, to future developments). And in the 1930s, after liberalizing its licensing policy and freeing itself from its corporate founders, it fought the charge that it was a "patent octopus" for pursuing litigation against weaker competitors and for dominating radio broadcasting. In the late 1930s, its bitter battles over FM radio broadcasting also cast the company, and Sarnoff himself, in an evil light. RCA archrivals Zenith Radio and Philco took advantage of those perceptions by attempting to alter (Philco) or escape (Zenith) RCA's television licensing suzerainty; Philco even accused RCA of attempting to extract information about its television research by spiriting some of its female employees out to Philadelphia nightclubs. RCA was eventually forced to cross-license its television patents extensively to all comers.

On August 10, 1937, Congressman W. D. MacFarlane, a Texas populist Democrat, rose in the House of Representatives to excoriate RCA in a typically populist critique. RCA, MacFarlane charged, had been born a monopolist, had persisted in that sin, and, in the face of a "notorious" government consent decree in 1932 that allowed RCA to operate despite an antitrust conviction, was now attempting to monopolize television as well. RCA's board, MacFarlane darkly claimed, was a nest of Morgan, Rockefeller, and other Republican interests (while CBS, RCA's broadcast rival, he added, was packed with bankers from Brown Brothers Harriman, Lehman Brothers, and W. E. Hutton & Co., a group with a slightly Democratic tinge). "Not content with their monopolistic control and the five percent on gross revenues they take from all licenses, they [RCA] began to terrorize even those who had license to compete against them," charged MacFarlane. In particular, RCA had forced smaller firms to wage costly patent battles, which these small companies, and lone inventors, could rarely win. "Such is the power of the patent racketeering of the Radio Trust."[8]

Ironically, by the time the war arrived, RCA's position in the radio marketplace was not as impregnable as its image as a radio trust suggested. Throughout the 1930s, the company had gradually lost market share to a dozen major radio rivals and another dozen smaller ones, all American companies. By 1940 RCA still led in global radio sales by dollar volume, but it trailed Philco in terms of worldwide units and, quite dramatically, by dollar sales in America. Hard on its heels were other RCA licensees, such as Zenith, Emerson, Galvin, Colonial, Belmont (soon to be bought by Raytheon), General Electric, Crosley, and Stewart Warner. In 1941 Zenith, too, slipped past RCA to take second place.[9]

Moreover, despite Sarnoff's best efforts and the complaints of regulators and competitors, RCA's patent fortress had also eroded by the early 1940s. A monochrome television receiving set is a complex electronic system—a cathode-ray tube with photoelectric cells and electron gun, intricate synchronization circuits, and banks of vacuum tubes. Despite its preeminence in television, RCA was still forced to cross-license for patents with Philco, Farnsworth, DuMont, and Hazeltine Electronics; Sarnoff broke one of his own sacrosanct rules by agreeing to pay royalties to the greatest threat, Farnsworth, in 1938.

Sarnoff had always poured profits back into R&D. As a result, RCA's dividends and earnings per share were small and its stock price—after the fantastic heights attained before 1929—was fairly low. RCA was never a world-class manufacturer in either quality or cost. General Electric and Westinghouse had done RCA's manufacturing for it well into the 1930s, and RCA lacked the flexibility to adjust to the shifting currents of consumer tastes. Only in 1929, when Sarnoff bought Camden, New Jersey's Victor Talking Machine Co., did RCA acquire manufacturing facilities and skills of its own. In the years before the breakup, Sarnoff also bought a General Electric vacuum-tube plant in Harrison, New Jersey, and a Westinghouse assembly plant in Indianapolis—both for stock. Nonetheless, by the end of the 1930s, RCA was still losing volume leadership in home radio.

Sarnoff attempted to forestall this kind of erosion in television. But there were more deep-set problems. The secretive, autocratic Sarnoff had never been able to build a strong management team; and by the time the war came, RCA lost the few solid managers it had. In 1943 he hired a merchandising executive named Frank Folsom—a veteran of the Montgomery Ward and Goldblatt department-store chains and a former chief of naval procurement—to provide day-to-day management and, particularly, to oversee television.[10] The navy, eager to upgrade RCA's manufacturing operations (particularly with Sarnoff himself away on war work—he served as a sort of public relations man and communications adviser to the military and would soon be "General" Sarnoff), "arranged" Folsom's availability. Unlike Sarnoff, who, while short and round, was a famously dapper dresser, Folsom shambled about like a rumpled bear. Folsom immediately plunged into the kind of detail Sarnoff ignored, preparing for the day when the FCC would approve television, getting RCA's manufacturing facilities in line, sounding the drum beats of retail promotion, lining up a national network of distributors and retailers, working out plans for servicing (early television was a far more troublesome product than was radio, leading Zenith's president, Eugene McDonald, to warn that it would just never work), and developing a grand strategy.

Folsom borrowed RCA's marketing strategy from the reigning king of American industry, General Motors chairman Alfred Sloan. It was a revealing choice. Early television sets were essentially bakelite boxes with bulging screens that were known as "doghouses." With the technology fixed and diffused across the industry, competition would take place on price and service and, most particularly, set design and screen size. Zenith early on tried to sell a round screen, arguing that since the camera lens and the human eye were both round, so should a television screen; the public rejected the analogy. (Zenith also unsuccessfully pioneered Phonovision, in which movies were sold over telephone lines, an early version of pay-per-view cable.)[11] RCA's basic model in 1947 had a square, ten-inch screen and sold for $375—a set that RCA, mixing its corporate metaphors, envisioned as its "Model T." The company had plans to develop a thirteen-inch screen, its Buick, and a sixteen-incher, its Cadillac.

Folsom also assembled a national service organization, running the risk of angering local retailers who had traditionally repaired radios. However, they were not unduly irritated, and early on, Folsom's servicing scheme proved a big success. In 1947 RCA was pleased to note that the average television required only six service calls a year, mostly for set adjustment and burned-out tubes.

In March 1947, when the FCC flashed the green light, Folsom's sales force of twenty-two hundred poured out to sell televisions. Pent-up demand was high, and RCA, its factories churning out five thousand sets a week, was limited only by the availability of the large kinescope picture tube blanks produced by Corning Glass Works and Owens Illinois Glass. RCA seized such a huge market share lead that by the summer, Folsom, fearful of antitrust problems and eager to broaden the market for NBC and pick up royalty income, invited competitors to RCA's Camden research laboratory to offer licenses on the technology. "Television is bigger than any one of us," he said, placing blueprints of the 630T—the Model T—on everyone's desks. By the end of the year, RCA had sold $40 million worth of television sets, more than the rest of the industry combined, and by 1949 the company had eighty-seven licensing agreements outstanding, which brought in an estimated $3.2 million. This figure would triple to over $9 million in 1950.

## The Revenge of the Marketers

RCA could not retain that dominance long. By 1948 Philco was leading the counterattack.[12] If RCA was the technical leader of radio, Philco was the marketing champ. The immediate predecessor to Philco had been founded in northeastern Philadelphia in 1909 to make storage batteries for cars—hence its name, the Philadelphia Storage Battery Co. In the 1930s, however, the storage battery business was undercut by the introduction of the RCA alternating-current tube, which eliminated many uses for storage batteries, particularly in radios. In response, Philco plunged into the radio business in 1928 by licensing the RCA patent package, and by 1930 it seized volume leadership. Although Philco offered a full line of radios—including car radios sold to Chrysler, Ford, Studebaker, and Packard—its greatest success came from its ability to sell low-priced, tabletop radios in tremendous volumes.

Despite its size and profitability, Philco was a much different company from RCA. Its headquarters, a barnlike building in working-class Kensington, provided a sharp contrast to RCA's Rockefeller Center headquarters. It had no prominent, visible symbol like Sarnoff. Its founder, James Skinner, retired in 1939 and was replaced by what *Fortune* called "a hard-hitting team" of long-time Philco executives led by President Lawrence "Larry" Gubb. "The executives sit around sides of a huge rectangular room. . . . No swank signalizes their getting together; they meet frequently in a board room that only recently has been prettied up to impress visitors, and more often than not they meet around one another's doorways." Philco seemed to have no greater ambitions than to make money and beat RCA; Sarnoff and Gubb particularly liked to snap at each other. Philco was quite content—even eager—to come into markets a half-step late, after the initial mistakes were made, wielding immense marketing clout. In 1940 it outspent RCA in radio advertising, $36 million to $22 million—and no one else came close.

Philco was characteristically cautious about taking on war work. Profits had been hurt by a strike in 1938, Skinner had retired in 1939, and its RCA suits were just being resolved by 1940. Despite its success, Philco badly needed a financial reorganization. The company had been privately held by about 120 shareholders, with stock used as bonuses to managers with the proviso that the company would buy it back upon either termination or death at 110 percent of the cost; the retirement of Skinner, the largest shareholder, had involved a big drain on capital. Philco's management was aging, and this policy posed a continuing threat to its working capital. In 1940, Philco went public, raising enough capital to retire an old preferred issue, eliminate its long-term debt, and bolster its capital.

What Philco really wanted from the government were radar contracts. Despite a prevalent view that the company was a second-class research outfit, Philco managed to obtain contracts to produce specialized radar and radio components, quartz crystals, cathode-ray tubes, and miniature vacuum tubes. Moreover, the company set up a school in Philadelphia for training servicemen in the maintenance and repair of electronic equipment (some twenty-five thousand passed through its classes during the war). Sales, nearly all to the government, soared from $77 million in 1941 to $153 million in 1944. Meanwhile, the company struggled to keep its all-important retail distribution network alive. Philco sent three managers crisscrossing the country, talking to manufacturers, scaring up any goods—from Fire King ovenproof glassware to Soil-Off cleaning fluid—that could be sold to its dealers to fill shelves and floor space. "We got them everything from cider to stockings," said one Philco executive.

Thus, after jousting with RCA for years, Philco had the cash, the name, the technical skills, and the distribution network necessary for the television offensive. Although it lacked the patent position of RCA, Philco did boast about its wide, flat screen, dubbed the Plane-O-Screen, and an ion trap in the tube that it claimed eliminated an unsightly brown spot. Still, Philco, like RCA, introduced basically the same product—a small monochrome screen encased in a big box that dominated a room. By 1948, with both Philco and General Electric picking up chunks of the market, RCA's market share fell to 30 percent and it looked as if the radio experience was about to repeat itself. In fact, it would repeat, but in a far more volatile fashion than radio had ever experienced.

In many ways, the leading edge of television in 1948 was neither the technological leader, RCA, nor the marketing leader, Philco, but a newcomer from Chicago called Admiral.[13] Its founder, Sicilian-born

Ross Siragusa—dark, handsome, with slicked-down wavy hair and a high-wattage smile—had already seen an earlier venture, a transformer company, fail in 1934, then formed a firm that looked cockily forward to television, Continental Radio & Television. In order to get an RCA radio manufacturing license, Continental had to agree to buy radios through a bankrupt RCA licensee, Radio Products Corp. In 1936 Siragusa, still only twenty-nine years old, bought the trade name Admiral for $100 and began selling cheap table radios, many of them for retailers such as B. F. Goodrich and Western Auto Supply. In 1941, he finally eked out a profit of $71,000, and a year later he merged Continental and Radio Products. Along the way he also managed to get a foothold into the appliance business, selling Flexo-Heat stoves and Dual Temp refrigerators.

Still, Admiral was hardly a household name. As late as 1941, it generated only $2.1 million in sales. Six years later it employed twenty-four hundred people, compared with Philco's fourteen thousand and RCA's forty thousand plus. Despite that growth, Admiral and Siragusa were still outsiders. Commander Eugene McDonald of competing Zenith Radio dismissed the subject of Admiral in *Fortune* in 1949. "He [McDonald] hasn't yet had the pleasure of meeting Admiral's President. As far as he knows, Admiral's President is the only important figure in the industry he hasn't yet had the pleasure of meeting."[14]

Admiral had no technical refinements, no patents, to bring to television. What it had, besides a distribution network Siragusa had built with wartime profits and two public offerings, was a willingness to break with the pack. After a short production run making large televisions, Admiral suddenly shifted to a new design, a smaller set called a consolette. "This new job, looking like a shrunken water cooler, will never be confused with a piece of furniture," wrote *Fortune*. "But at least it stands quietly in the corner without hogging the living room table."[15] Admiral priced it at $330 and it sold and sold. Soon afterward, the company introduced a matching cabinet that contained an AM-FM radio, a phonograph, and space for a television: an integrated home entertainment system. Admiral followed that up with a consolette with a larger screen that sold for $80 less than the original. Along the way, Siragusa spent nearly $2 million advertising on television alone, including sponsoring one of the first televised professional football games.[16] In 1949 Admiral's revenues more than doubled, to $112 million, and the company trailed only Philco and RCA in market share. In terms of how effectively it used its assets, Admiral ran away with the game, racking up a 27 percent return on assets to RCA's 9.6 percent and Philco's 9.5 percent.

Admiral's success was one sign of what really mattered in television—not innovative technology, which in its gross features was fixed, but the *perception* of technology. While RCA was boasting about new models with fewer tubes, Admiral was quick to market sets that displayed a sort of crowd-pleasing, ersatz technical sophistication and elegance: one of Admiral's more expensive console sets ($775) featured a "Magic Mirror" television, a "Dynamagic" radio, gold-plated control knobs, and an on-off "jewel light." This was a mass consumer audience, most of whom would not know a thermionic tube from a Dynamagic radio. Moreover, Admiral also began underpricing, putting pressure on margins across the industry, while continuing to fuss over Siragusa's greatest creation, the distribution system (Admiral eventually boasted sixteen thousand outlets), and cranking out sets at its Chicago factory, which, it claimed, had the longest assembly line in the industry. Mostly because of Admiral, RCA found its share skidding as low as 12 percent in 1950.[17]

## Boom and Bust

The early days of television established a pattern that would be repeated in mass consumer electronics from tape recorders to Walkmen to videocassette recorders. Pioneering firms found that achieving further advancement was like slogging in deep mud. Either the technology was fixed by regulation—and patents were widely diffused—or new technologies quickly leapfrogged the old. In both cases, pioneering firms found that they would seize large market shares only to have few effective means, short of quickly racing ahead technologically (the Sarnoff solution), to defend their positions. All consumer marketers faced a dilemma: Should they pour money into fundamental research in the hopes of getting into a market early? Or should they wait, husbanding their resources for a marketing push? Despite the talk of technological revolutions, consumer electronics—televisions, VCRs, even personal computers—remained, as the situation had been in radio, a game of cost (manufacturing) and image (marketing) over advanced research and development.

Television quickly settled into cycles of boom and bust characteristic of commodity products. In the early 1950s, demand suddenly sagged as the first wave of buyers petered out and, with talk of color in the air, new buyers hesitated. Set sales hit 7.5 million in 1950 but then fell back to 5.6 million in 1951. Companies were caught holding large inventories and struggling to fund marketing campaigns. Raytheon's Belmont, for instance, had failed to assemble a top-flight distribution network and was forced to sell to outlets rejected by the majors and to a few big, tight-fisted retailers such as Montgomery Ward. With the start of the Korean War in 1952, Belmont, like a number of other manufacturers, was hard hit by new restrictive credit rules and began to liquidate its inventory, borrowing from its parent to stay afloat.[18] By 1953, Belmont owed parent Raytheon $11.5 million. Raytheon's chairman, Charles Adams, sent a Raytheon senior executive, Harry Argento, to reorganize the operation, while ordering a crash effort at home to develop a color television tube. When Argento got to Chicago, he discovered "a real mess. It [Belmont] was running a faulty assembly line, billing single orders at a time with conflicting orders being issued; the employees were engaging in wholesale thefts and there were leakages in inventory. Great losses were piling up."

Meanwhile, Raytheon, still with its belief in a technological manifest destiny, was painfully discovering the realities of television. "As an engineer," said Sidney Standing, a Raytheon television engineer, "I deplored the switch of emphasis dictated by the merchandisers that changed television from a great technological search into a piece of furniture. It slowed the rate of progress and switched the emphasis of the state of the art." In fact, the fault was not with the "merchandisers" but with the television standards policed by the FCC (Raytheon received most of its revenues from a different wing of the government, the military, which set similar nonmarket, if often more technologically challenging, specifications). Raytheon developed what it thought were technologically advanced models for Belmont—the Challenger and Aristocrat lines of 1954—and was working on a color tube and new germanium diodes; but the losses mounted. In 1955 Adams killed off the Raytheon Distributing Company, Belmont's marketing arm. "It lost money for every month of its existence except for the first few," he said, "and those losses far exceeded its capital investment. It didn't even pay for the overhead in the offices." By then Adams was looking for the exit. Finally, in 1957, he announced that Belmont would be sold to the king of the merchandisers, Admiral. Siragusa, from a position of strength, negotiated a tough deal. Raytheon had to fill its outstanding orders from Montgomery Ward and pay down the inventory. Admiral then brutally eliminated staff; Argento ended up at Philco.

Belmont was not an exception. The problem for many struggling television makers was not in gaining access to technology but in selling enough sets to support future investment. Costs rose as companies were faced with gearing up for color television in the 1950s. And with consumers uncertain about the advent of color, companies faced a double marketing challenge: preserving monochrome sales while talking up color. In 1957, total set sales sagged to 6.5 million from 7.4 million in 1954, the year color was first approved for sale by the FCC.[19] Inventories again rose, the average set price slipped from $225 to $190, and margins tightened. By then, consoles were being eaten up by even smaller sets, called portables, first introduced by General Electric. A shakeout began. Within the year, Magnavox acquired Spartan Corp.'s television unit and Sentinel Radio Corp.; Stromberg-Carlson, a famous radio name acquired by General Dynamics, dropped out of television; CBS-Columbia, a unit put together by Paley to push mechanical color television, also quit (claiming that Korean War shortages had done it in), as did Bendix Radio, Crosley, and the Bendix division of Avco Manufacturing. All in all, over one hundred companies left the business. As they went, they dumped inventory, further driving prices down. Color produced an industry with fewer, larger companies—despite an antitrust policy that sought to encourage many smaller companies to compete. The industry advanced up a series of terraces: first black-and-white, then color, then the slippery, precipitous staircase of solid state. Technological costs rose with each step; RCA spent considerably more to develop the more complex color tube than it had spent on black-and-white. At each stage, more companies dropped out, expiring from either financial weakness or marketing debility.

Color itself almost died in the late 1950s. Most of the firms selling color sets had to buy RCA's round, shadow-mask picture tube. The technology was still imperfect—color sets were particularly maddening to adjust—and, except on NBC, there were few color shows to watch. In the late 1950s, manufacturers began to drop out of color. Once again, RCA was hit by litigation from Philco and Zenith and a federal antitrust suit, this time with criminal penalties. Sarnoff eventually agreed to a public pool of one hundred color television patents and paid a small fine. Sales only slowly revived in 1961 when Zenith, by then number two to RCA, began making its own color tubes and CBS and ABC finally began the long-awaited shift to color broadcasting. Three years later, Motorola pioneered a shorter, rectangular color tube that proved very successful, allowing it to move to number three. RCA and other tube makers soon went rectangular, phasing out the old round tubes. By the mid-1960s the mass shift to color had begun.[20]

The transition to color allowed RCA to regain market share—it hit 70 percent of (color) sales briefly in 1962 before dropping back—but shook up the industry.[21] Both Admiral and Philco suffered. In 1957, James Skinner, Jr., the forty-two-year-old son of the man who built Philco, took the helm of the Philadelphia company, which, becalmed in the late-1950s downturn in consumer spending, was adrift.[22] Television sales in particular were lagging. Philco had never won the leading role it had in radio, and, while it was still third in 1959, its profitability, never high, was sinking further. The dividend had been eliminated after a big loss in 1956 and margins had slipped below one percent. Besides paying for the ongoing costs of television, Philco was also pouring capital into a transistor operation in Lansdale, Pennsylvania, and a new computer project. Tough and caustic, Skinner, with his GI crew cut, immediately launched a deep cost-cutting campaign, forcing much of senior management into retirement, followed by a massive reorganization designed by the consulting firm of Arthur D. Little.

Two years later, Philco's television share had fallen further, the dividend had not been resumed, the transistor and computer projects were consuming cash, and the company was spending money on a laboratory in Palo Alto to develop ground control equipment for space vehicles. In the first half of 1961, Philco lost $8.4 million because of weakness in semiconductor pricing and appliance sales. Enter Ford Motor Company, which in September 1961 offered to buy Philco in a stock swap, a deal worth about $100 million, plus the assumption of $44 million in long-term debt, the first major acquisition by the carmaker's young chairman, Henry Ford II. Skinner and the Philco board accepted, talking enthusiastically about tapping Ford's deep pockets. The press agreed that Ford was eager to get into the appliance business like its rival General Motors, and that Philco's defense electronics business could help Ford's small missile and defense electronics division. One observer even compared the acquisition with Du Pont's rescue of General Motors after World War I, arguing that "Ford . . . [was] loaded with management and marketing potential" and, like Du Pont, needed a place to put them.[23] This, of course, is a misreading not only of the Du Pont–GM alliance but of the future Philco–Ford relationship and of Ford's own capabilities.

Philco never regained momentum. Ford did not prove to be an inspired consumer marketer, and television executives found it difficult to compete with other uses for Ford's large, but hardly boundless, supply of investment capital. Television costs were rising and returns were not. Television had become that terrible paradox: a high-cost, technologically complex commodity product. Philco struggled along with less than 5 percent of the color market. By 1973 Ford was weary of the game. From July to September of that year, every integrated circuit Philco bought from its supplier, Motorola, turned out to be defective—which Philco failed to discover until the sets were assembled.[24] The cost of that debacle convinced the automaker that television was more trouble than it was worth. Ford quickly sold Philco's color TV tube plant in Lansdale to Zenith; then, in October 1974, it unloaded its entire U.S. and Canadian home entertainment business on Sylvania, a subsidiary of GTE. A few months later, Ford also sold to Sylvania Philco's Taiwan television unit, effectively exiting the business.

## The Shift to Solid State

The transistor, a semiconductor component like the crystal, was invented in 1948. By the early 1950s it was beginning to replace vacuum tubes in devices like hearing aids. By the later 1950s, engineers had begun to solder transistors together on circuit boards that could replace large numbers of vacuum tubes in devices such as radios and televisions; such circuit boards were the predecessors to the integrated circuitry that would begin to appear in the 1960s. These so-called solid-state components dramatically reduced television service problems, power requirements, warmup time, and the heat radiated by sets with vacuum tubes; they also allowed smaller, lighter sets to be made. With solid state, television manufacturing changed from an industry faced with an occasional wracking metamorphosis to one trying to cope with continuous improvement. On the production side, television began to resemble less the auto companies and more the semiconductor firms. Solid state forced companies to make continual, incremental improvements, not just to boost set quality but to drive down manufacturing costs.

The appearance of these so-called solid-state systems accelerated the rate of change within the rigid frame of television technology—while applying pressure to that frame. When transistors began to appear in televisions, the underlying basis of the industry changed. In many ways, it was a more fundamental shift than the transition from monochrome to color. But it was silent—and seemingly invisible to much of the U.S. industry.

Television in the United States was certainly thought to be a progressive, productive, modern industry. From 1958 to 1966, the U.S. Bureau of Labor Statistics estimated that output increased 222 percent in television, while employee man-hours grew only 97 percent for an annual gain of 6.2 percent, more than twice the yearly productivity gains of the economy as a whole.[25] In 1958 the industry was spending only $200 in new capital per employee, a third of the average for all American manufacturing; by 1966 that had grown to $930, just below the average.

However, while microelectronics and automated testing did boost productivity, the Bureau of Labor Statistics still predicted that the long assembly lines manned by large labor forces would persist. While solid state replaced vacuum tubes within the television box, it did not replace the largest tube of all—the picture tube.

The shift to solid state began in the mid-1960s, at the peak of the color boom. As in black-and-white, color began with large, furniture-like sets. In 1960 Sony, a Japanese firm best known for making "transistor" radios, introduced the first all-transistor television. Soon, Sears, unable to strike a manufacturing deal with RCA, began buying large numbers of transistorized portables made by Japan's Sanyo; other large American retailers quickly struck similar deals with eager Japanese producers. More characteristic of American manufacturers, however, was Ford-Philco's strategy. It sold a solid-state "portable" that cost $795 and weighed "only" eighty-five pounds; it still had some tubes, which Philco explained by declaring that it was "transistorized where it needs it most."[26] Although Fairchild Semiconductor would build a demonstration solid-state television in the mid-1960s in the hopes of spurring chip sales, U.S. television companies would not completely embrace solid state for another decade; instead, they would bicker about the merits of "hand wiring" (Zenith) over printed circuit boards (RCA). Even when American firms did move toward extensive use of multiple circuit boards, Japanese manufacturers stayed ahead by integrating its components onto *one* board. As late as the 1970s, RCA would invest heavily in a scheme for embedding components in a ceramic base—a move that wasted time and money for RCA; ceramic could not compete with integrated circuitry. Still, as late as 1970 little had seemingly changed. RCA still had a dominant 23 percent of the color television market, with Zenith (the leader in black-and-white) in second with 21 percent and Motorola in third with 6 percent. American firms controlled well over 90 percent of the market.[27]

The black-and-white survivors had thrived in the 1960s by feasting off color ("Without color TV," snapped Sarnoff, "they'd be out on the corner eating a ham sandwich."), predicting that they would never repeat the mistakes they had made in monochrome. Of course, they did repeat them. By 1968 the industry had the capacity for ten million color sets but sold only six million. Color tube prices remained high, labor and development costs were rising, and the red ink flowed. Share prices fell, making financing more expensive and providing a rationale for exiting the business. "The money and technology required to stay ahead in color television is incomparably greater than black and white," lamented

Joseph Wright, Zenith's president in 1969. "In black and white you had companies in the loft turning out sets and selling them almost by intuition while they benefited while someone reduced the cost of components. With color, you just can't survive that way."[28]

Admiral, still controlled by Siragusa, had managed to survive, although the stock never approached the $130-a-share high of 1950. Like a paper airplane on a windy day, Admiral soared in both black-and-white and color sellers' markets, only to be slammed to earth by high costs and large inventories when the bust came.[29] In 1967 the company took its first yearly loss. In 1969 Siragusa's wife died after a long illness; soon afterward, he appointed his son, Ross, Jr., as president and chief operating officer, just in time to absorb a big operating loss in 1970. Like Skinner, Jr., Siragusa, Jr., was forced to cut costs, consolidating and closing plants and selling off the company's tube operation to RCA. By the early 1970s, Admiral had slipped to sixth in market share; in 1972 it raised money by selling equity, diluting the family stake.

In 1973, profits revived. In September, Siragusa, sixty-six years old, married a second time in a splashy wedding at the Vatican that included a visit with the pope at Castel Gandolfo. The newlyweds then made their way home after checking out a new yacht that Siragusa had built in the Netherlands.[30] Back in Chicago, Ross, Jr., was negotiating an exit for the family. Rockwell International, the large space and defense contractor, was nervous about a downturn in defense spending as the Vietnam War wound down and wanted to expand its consumer businesses. Horizontal mergers were frowned on by the Federal Trade Commission. The multi-industry corporation, or conglomerate, was on the rise, and Rockwell drifted in that direction. Like Philco, Admiral needed deeper pockets; merger negotiations began. A figure of $100 million was quickly reached only for Admiral to discover sales and profits sliding again. Profits thudded 81 percent in the fourth quarter of 1973, and net income fell 27 percent for the year. By the time the final papers were signed, Rockwell had acquired Admiral for $78 million, allowing the Siragusa family to take home about $25 million.

Once again, Wall Street and the press praised the move. "Analysts asked agree that Admiral has basic strengths and can fill the 'gap' in Rockwell's business," said the *Chicago Tribune*.[31] Some gap. Admiral quickly plunged into unprofitability. In December 1973 Rockwell quietly sold off Admiral's Taiwan TV plant. As at Raytheon and Belmont, defense-oriented engineers at Rockwell produced solid-state designs for Admiral televisions that were incompatible with the realities of a consumer market like television; friction between the two groups sprang up.

Four years later, Rockwell sold Admiral's U.S. home appliance business, Magic Chef, for $62 million, followed three months later by the sale of Admiral Corp. of Canada for $34 million more. Rockwell blamed Japanese competition for the demise of Admiral.

## Decay and Collapse

After the scramble for color, a certain stability seemed to return to the television set industry. Foreign competition was viewed as a serious, though not lethal, problem (Zenith won a charge of dumping against Japanese firms in 1971, though the Treasury failed to act), and new domestic threats simply did not exist; barriers to entry were too high, returns were too low. Then, beginning in the mid-1970s, it all came apart: in the space of a few years, Admiral and Philco were broken up, Magnavox and Sylvania were acquired by the Dutch electronics giant Philips, and Motorola's TV business was bought by Matsushita. By the end of the 1970s, all the Japanese firms were making televisions in North America and Korean and Taiwanese makers were following close behind.

Japanese firms, most of which had openly licensed their technology from RCA, had begun by selling smaller, low-priced sets in the 1960s. Sony led the way technologically, first with its solid-state portables, then, in 1968, with the Trinitron color tube. Part of the reason for the Japanese firms' intense export push was a consumer market at home that developed slowly and intense competition among a large number of well-capitalized companies. By 1970 all the Japanese exporters had gone 100 percent solid state, a boast only Motorola, which was also in the semiconductor business, could make among U.S. companies. The powerful move into solid state had forced the Japanese to install automatic insertion equipment in their assembly plants—a process first developed in the United States to make car radios in the 1950s—thus slashing labor and component costs.[32]

American firms, however, lagged in understanding the implications of solid state, particularly on circuit board design, and they stuck to hybrid sets and manual assembly. The threat of trade restraints, which culminated in the Orderly Market Act of 1977 limiting Japanese television imports, induced Japanese companies either to ship through Taiwan and Korea—building up industries in those countries—or to manufacture in America, ironically at the same time the American firms were bemoaning high labor costs and moving plants to low-wage regions, such as Mexico and Brazil. The tide turned swiftly. In a 1968

*Consumer Reports* survey of color television, none of the eight sets tested was foreign made; ten years later, only five of fifteen sets were made in the United States. In 1974, Matsushita (with its Panasonic, National, Victor, and, from Motorola, Quasar labels), passed RCA as the largest television producer in the world. The Japanese also dominated in terms of quality, a direct result of automation and extensive pretesting of components (a process that would have quickly revealed Philco's integrated-circuit problem). In a 1974 evaluation of eleven sets, nine of which were American, *Consumer Reports* complained that "five sets required the attention of a repairman right at the start. . . . We had to give up the idea of testing the twelfth, Admiral. Each of the three Admirals reached us in disabled shape." Throughout this period, the ostensible technological leader, also RCA, lags in the quality rankings.[33]

Throughout the charges and countercharges over trade, RCA remained aloof. It had licensed its technology to Matsushita early on, and it earned a considerable royalty from Japanese triumphs. Moreover, it continued to own a major share of the market. On January 1, 1965, the General, then seventy-five, finally gave up operating responsibilities, making his son, Robert "Bobby" Sarnoff, president and RCA's former technical chief, Elmer Engstrom, chief executive officer.[34] Robert inherited a company in transition. The General had made one last attempt to reprise radio, as well as black-and-white and color television. In the late 1950s, the General began investing in an RCA mainframe computer project. However, he underestimated the enormity of the investment required in computers and the strength of the dominant competitor, IBM (a "new" RCA, right down to the antitrust charges, though one that was far more profitable than RCA ever was). RCA's efforts were dogged by bad luck and poor decisions. Both Sarnoffs starved further television research to pay for the computer initiative, inviting competition by licensing television patents freely overseas and keeping prices up. And yet they did not spend nearly enough to challenge IBM effectively. The General had also launched RCA's conglomeratization drive with the acquisition of appliance maker Whirlpool; Robert continued it by purchasing publisher Random House, Hertz (the rental car firm), a New York real estate firm, a rug manufacturer, and a chicken processor. Some of these units, such as Hertz, were also voracious competitors for capital with the computer project.

Robert Sarnoff lacked the vision, the drive, and the megalomania of the old man. Father and son fought, particularly over cosmetic changes—such as Robert's decision to update the traditional RCA logo and to abandon the company's formal name, Radio Corporation of America, for a more "modern" RCA (the General lost the redesign battle but temporarily won the name change). RCA was soon under serious attack on a number of fronts, particularly as the recession of the early 1970s commenced. In computers, the company picked up share but suffered mounting losses. The television tube industry was plagued by overcapacity. RCA's television share was under renewed attack by Zenith, Motorola, and Magnavox. Semiconductors were losing money. CBS was pounding NBC in broadcasting. Avis was picking up share from Hertz. And RCA's defense electronics business was slipping. Finally, in September 1971, Robert admitted defeat in the computer wars and announced a humiliating $490 million write-off. "It was the greatest disaster since the Ford Edsel," writes Robert Sobel in his history of RCA.[35] The General, failing for some time, was confined to a bed in his East Seventy-first Street townhouse apartment. On December 12 he died in his sleep.

RCA never recovered. Four years later, with earnings slumping again, the RCA board fired Robert. Not one of his three successors was able to salvage the company, despite extensive asset sales and restructurings. RCA's last hurrah was a major effort in laser videodisc technology for television. After initial promise, RCA's elegant technology was buried by the popularity of predominantly Japanese-made videocassettes; RCA pulled out after $500 million in losses. Finally, on December 12, 1985, RCA's last chairman, Thornton Bradshaw, announced that he had sold the company to one of its original parents and fiercest competitors, General Electric, for $6.28 billion. It was a deal made possible only by the relaxation of antitrust enforcement by the Reagan administration. Within the next year, GE sold the David Sarnoff Laboratories, where television was first commercialized, to the Stanford Research Institute, and it sold RCA's television and audio equipment business to Thomson, the state-owned French electronics company.

Who was left? Ironically, only Zenith—one of RCA's oldest and most litigious rivals (the General once called Zenith's McDonald "that parasite"). By the late 1980s, Zenith, long the least diversified of U.S. television makers and the only one to focus strictly on the domestic market (one reason for its willingness to sue the Japanese), was barely able to hold on in television, against the wishes of much of Wall Street. And even Zenith had its circuit boards assembled in Mexico.

## Behind the Fall

What befell American television? Explanations for its demise are legion: entrepreneurial decay; a failure to reinvest, particularly in manufacturing; an emphasis on marketing over manufacturing; the recovery of Japanese and German industry; foreign dumping. Each of those factors played a role, yet several further points need to be made here. While the Japanese inflicted the damage, they were attacking an industry ready to succumb. Twenty-five years of commodity competition did not leave it hardened by competitive fire, as conventional business wisdom would have it; instead, it was left weaker and more fragile, its coffers bare, its management discouraged, its friends few. The busts, the shakeouts, the fierce rivalries, the prospect of continual competition, the inherent limitations of the technology—all combined to weaken its will, like a boxer who has taken too many shots to the belly. Wall Street was not optimistic about television, and neither were many of television's own managers, who cut back on R&D after 1975. Even RCA and Zenith, the two market leaders of the 1970s, tried to diversify rather than reinvest their capital back into television. Zenith became, for a time, a respectable producer of laptop computers; RCA disastrously conglomerated.

Just as important, the industry failed to recognize the nature of the technological change that overtook it with the rise of solid state. Television managers were blind to the effect of the transistor and, later, the integrated circuit and microprocessor, which forced continuous and accelerating change upon it; and they were slow to realize that manufacturing could no longer be separated from design or marketing. Television was a branch of electronics, but one that harkened back to the scale economies of the radio age. The advent of the transistor forced its face toward an unsettling future.

But television was literally trapped in its past. By regulating standards and opening access to patents, policymakers inadvertently created the commodity competition that followed. Normally, such competition would produce oligopoly. Thus, particularly with the advent of solid state, which initially forced capital costs higher, an oligopoly had formed in the United States by the early 1970s. Enter then the free market, not only in the form of international trade, which brought the Japanese, clutching their RCA licenses and their Sears orders, but in the form of capital markets, which rejected the commodity returns generated by television manufacturers and influenced corporate executives not to reinvest their capital in a mature business line. Television was a discouraging business by the 1970s.

And yet it is a mistake to assume that the inner dynamics of television, or any technology, would remain static. The recent scramble to develop high-definition television (HDTV) represents the apotheosis of solid state, in a television context.[36] For all its disruptive effects, the advent of solid state did not transform the structure of television manufacturing or overthrow scale economies; it simply allowed one group of oligopolistic producers to crowd out another. HDTV, on the other hand, represents the point where accelerating change produces the kind of transformation that may actually force open the closed world of television manufacturing to new competitors, albeit for only a short time. With its larger screen and significantly brighter, sharper images, HDTV represents the first fundamental shift in television technology since the advent of color; in fact, it could turn out to be the most radical shift since the 1930s. The Japanese were quick to see its promise and perils. After developing their own version (research in Japan on HDTV began as long ago as 1965), the Japanese companies, particularly Sony and Hitachi, seemed to have a lock on the television future.

However, that no longer appears to be the case—ironically, because of the temporary lifting of many of those same limitations that made television a conservative technology in the first place. First, the Japanese firms found their long technological lead reduced when they became mired in a global struggle over standards in both America and Europe. Those delays, like those that beset early American television (and reduced RCA's lead), sprang both from the desire of broadcasters to put off any large capital investment and from non-Japanese governments and corporations eager to slow the Japanese juggernaut. Second, those delays have allowed competitors to master leapfrogging technologies while the regulatory environment is still fluid. These new technologies, which hinge on the use of digital signal processing as opposed to traditional analogue, have their genesis in the computer and semiconductor industries, which in the American sphere are characterized by intense competition among many companies, large and small, fed by capital raised on Wall Street. Zenith, for instance, has managed to transfer its digital skills acquired in its laptop computer business to a promising digital HDTV system (ironically, to pay for the effort, Zenith had to sell off its laptop business).

No matter how the HDTV wars are resolved, the current fluidity in HDTV will last only as long as the forces of regulation fail to set standards. Once those standards crystallize worldwide, HDTV will be placed in a conservative harness again, technological advance will proceed down a fairly narrow track (particularly as millions of consumers build an installed base), and a fairly small number of large companies will inevitably exploit the technology as long as over-the-air television remains a central part of home entertainment systems. Regulation will not countenance continuous fundamental innovation. Therefore, companies around the globe must argue for standards that play to their strengths.

There is a final lesson here that applies to consumer markets in general. All technologies eventually mature; consumer technologies, like television, tend to mature more quickly. Why? First, consumers represent a potent political force that regulatory agencies labor to keep quiescent and satisfied. Regulation is thus designed to protect, or insulate, consumers from the treacherousness of the market and the vagaries of technological change. But even without government regulation, consumer markets tend toward conservatism over the long term. Mass consumer markets resemble mass democracies: buyers vote with their wallets, and they vote decisively. Majority tends to rule. Once consumers agree on a standard, they are not eager to change, to toss out their hard-earned goods and pioneer something truly novel, to risk buying a round screen in a world of square screens or a laser disc in a world of videocassettes. As a result, manufacturers, too, grow wary of too much innovation, too much revolution, and pragmatically embrace what has worked in the past, while, like David Sarnoff, talking grandly of the future.

# 5

## Centrifugal Tendencies

### Structures of Innovation

For most observers, the early television wars—RCA versus the world—fit into traditional patterns of behavior and thought. For New Deal liberals, the Bible of economic orthodoxy was *The Modern Corporation and Private Property*, an analytical look at American big business by attorney Adolph Berle and economist Gardiner Means. When the book appeared in 1932, the Great Depression was near its nadir and the book's description of corporate concentration—particularly its charge that two hundred American corporations controlled nearly half of America's non-banking corporate wealth—found a large and sympathetic audience. Berle pounded home the point that such a concentration was politically and economically damaging, and he called on the federal government to alleviate the imbalance. As the 1930s wore on, the fear of monopoly produced New Deal initiatives that revived the energies of the Progressive Era trust busters: a new emphasis on aggressive antitrust policies under U.S. attorney general Thurman Arnold; increasing attention paid to small business, particularly at the hearings held in the late 1930s by the Temporary National Economic Committee (TNEC), which through thirty-three thousand pages of testimony in forty-five volumes ("a miasma of words," wrote John Kenneth Galbraith) labored to expand the case made by Berle and Means; and much talk of refashioning patent laws and developing new financing sources.

Among economists, technological innovation was invariably viewed in structural terms, when it was discussed at all. What form of economic structure could best generate innovation? There were two schools of thought. First, there was the view shared by Midwestern populists who had battled against the encroachment of "big business" since the days of William Jennings Bryan, and by followers of Supreme

Court justice Louis Brandeis who championed small, atomized, decentralized business units. Both groups charged that small businesses (and lone inventors) had been driven from the scene through corporate exploitation of economic weapons such as monopoly pricing, restraint of trade, patent encroachment, and litigation; inventors like Farnsworth and Armstrong were often pointed to as prime victims. Brandeis in particular argued that large corporations manifested an unnatural, hence inefficient, growth. Frank Dunstone Graham, a Princeton University economist, provided details to this indictment in his 1942 study, *Social Goals and Economic Institutions:*

> The large corporations do not take great chances in putting new and improved goods on the market. On the contrary, they tend to engross inventions, and to retard their appearance, until it is all but certain that the new product can be marketed under depressed general economic conditions. This is very far from providing an adequate amount of enterprise. The large corporation is, in any event, not well adapted to new enterprise. . . . Quick decisions are difficult; new proposals have to run a gauntlet of critical investigation by ubiquitous vice presidents; red tape is prevalent. A new and comparatively small concern, under owner-management, would ordinarily be a much better medium for the launching of innovation.[1]

Even I. F. Stone echoed the Brandeisian thesis when he wrote in 1944 that the "special interest" of Kilgore's war mobilization committee "has been in our technological resources, that is, in our brains shackled by monopoly."[2] Such a message, with its combination of economic and political argument, would recur in the decades ahead. In theory, it was an argument that could be shared by both liberals eager to break the power of big business and conservatives intent on restoring some semblance of perfect competition.

But by 1945 this attack on big business was waning. Like the World of Tomorrow, with its bow to science and its eager display of technology, the New Deal, particularly in its later stages, heaped praise upon small businesses and intrepid inventors but accepted the inevitability of the giant corporation and its industrial R&D laboratories. The war had been fought and won with a strong central government and large units of innovation and production; central planning was in vogue, and Keynesian economics, with its emphasis on manipulating macroeconomic aggregates, was merging with cries for full employment and growth. The TNEC, set up to probe monopolistic behavior, instead painted a much more complex picture in which industrial economies of scale were portrayed as not only productive in many cases but, in an economy based on industrial technologies, inevitable. And Thurman Arnold's antitrust efforts, activist to be sure, appeared less intent on returning the economy to a world of small units as on supervising big business. Among mainstream economists, sheer size appeared to possess unbeatable advantages.

Economist Joseph Schumpeter defined what appeared to be the cold, hard realpolitik of innovation. Nearly sixty years old, he had returned to a subject, the entrepreneur, that he had first pioneered thirty years earlier. His *Capitalism, Socialism and Democracy*, published in 1942 (republished to greater note in 1947), elaborated on a scheme of a dynamic capitalism periodically transformed by the entrepreneur. Unlike many of his future disciples, Schumpeter was not an unabashed admirer of small business; his thinking had moved from the powers of the "exceptional entrepreneur" to the large corporation, with its R&D capabilities and some measure of market control, which could internalize "gales of creative destruction." He even saw much good in the monopoly. "As soon as we go into details and inquire into the individual items in which progress was most conspicuous, the trail leads not to the doors of those firms that work under conditions of comparatively free competition but precisely to the doors of large concerns. . . . There are superior methods, available to the monopolist that are not available at all to a crowd of competitors."[3] He particularly emphasized the important role that financial strength played in pursuing new innovations and the advantage in terms of long-range planning that monopoly pricing allows. And he attacked the waste and "bacilli of depression" of perfectly competitive industries.[4]

How to reconcile the entrepreneur with the monopolist? Schumpeter's entrepreneurs were agents achieving new economic combinations, creating a new, higher economic equilibrium. An entrepreneur could be a manager, an owner, a major shareholder. Entrepreneurship was determined less by economic structure than by a quality of the spirit. And, indeed, Schumpeter's larger message in *Capitalism, Socialism and Democracy* is that contradictions in capitalism lead to a dampening of that spirit just as capitalism achieves its greatest success, ushering in its decline and replacement by socialism.

How did Schumpeter come to what seems today such a paradoxical argument, with its deification of the exceptional entrepreneur? The evidence was all around him. For example, although hundreds of new radio manufacturers were established in the 1920s, only a handful conducted any radio research at all; and in a number of those years, more companies failed than were formed. Sarnoff's RCA, on the other hand, poured $50 million into television development alone, probably more than the rest of the industry combined. "I want you to copy RCA's product so completely in every detail, even though you believe that you can improve on some of its features," a radio executive reported he was told by his supervisor in the 1920s. "When we have learned to do it successfully, then we can think about improvements."[5]

AT&T's Bell Laboratories was an even brighter example of what wonders monopoly power could spawn. American Bell Telephone (AT&T was originally its long-distance arm) had long used technological mastery as a weapon. In the late nineteenth century, American Bell accepted the separation of R&D functions: universities performed pure research; corporations developed patents from independent inventors; and the government stayed away, except in wartime.[6] The company rethought these tactics as the demands of telephony, particularly long distance, forced it to act more proactively. Bell had covered all the angles in 1900 by buying competing patents for a loading coil transformer—a device that refreshed telephone signals, thus allowing for longer-distanced service—from both Columbia University professor Michael Pupin and one of its own researchers, George Campbell. When the inevitable interference suit surfaced, Bell attorneys defended Campbell (incompetently, as it turned out) and Pupin won the case. It was a classic example of patent monopoly: AT&T used its financial might to lock up both alternatives of a development that reaped millions of dollars, thus gaining a devastating edge over smaller competitors.

The next step, however, required something more than a patent deal. At the turn of the century, J. P. Morgan had reorganized the company after the panic of 1903, moving its headquarters from Boston to New York and setting up AT&T as the holding company. By then it was clear that the loading coil could extend transmission distance only so far. In 1907 de Forest invented his audion, and researchers at AT&T quickly saw that it might be adapted to transmit rather than receive, further extending the telephone's reach. Frank Jewett, the young assistant to AT&T's chief engineer and an electrical engineer who had studied physics under University of Chicago physicist A. A. Michelson, suggested that "real" scientific advice be sought. Jewett wrote to Chicago physicist Robert Millikan, a junior colleague of Michelson who was investigating electrical discharges, requesting that he send AT&T several of his best students. In 1911 the company formally set up a research unit under Jewett's direction, with Millikan as consultant. "A subtle change had occurred," writes Thomas Hughes in his history of American technology, *American Genesis*. "Instead of talking of inventing an amplifier, as would have been common earlier, now they [at AT&T] anticipated applicable discoveries flowing from fundamental research."[7]

Harold Arnold, the first of Millikan's boys sent to Jewett, made the key breakthrough in the rigorously rational development of the audion by evacuating air from it to create a true vacuum tube. The successful development of this device, the repeater, allowed AT&T to offer transcontinental telephone service, which it inaugurated, with great hoopla, at the Panama-Pacific exhibition in San Francisco in 1915. AT&T's control over the audion gave it a central role in the early radio industry (and a stake in RCA, as well). A decade later, AT&T's various research units employed some three thousand employees and boasted a budget of $12 million. The company consolidated those units, setting up Bell Telephone Laboratories with Jewett as its president. Two years later, Clinton Davisson, with assistant Lester Germer, serendipitously discovered proof of the wave nature of the electron, then a problematic hypothesis in quantum physics, while working on vacuum tubes at Bell Labs. A decade later, Davisson won the company's first Nobel Prize.

By World War II, Bell Laboratories was recognized as a national resource (the company tried, and failed, to kill an antitrust probe by claiming that it was indispensable to national security) and the most obvious proof of the Schumpeterian hypothesis equating size, monopoly power, and innovation. With assets of $5 billion, AT&T was by far America's largest corporation. It controlled 83 percent of all telephones and, through Western Electric, made 90 percent of all telephone equipment.[8] It also had more shareholders (637,000) than any other company, which in turn brought it under the guns of Berle and Means, who contended that wide shareholdership created a situation in which management was beholden to no one.

Bell used its monopoly profits to build a laboratory that even won respect in scientific circles. In many ways, Bell Labs was more hospitable to scientific research than were the elite American universities, not to mention other corporations. There were no teaching responsibilities, no need to raise money, no faculty committees to sit on, and salaries were generally higher than in academia. There was considerable interdisciplinary interaction; in fact, fitting into the "team"—a reflection of a technically complex telephone network—was viewed as the key to success at Bell Laboratories. Turnover was rare; there were few places to move to, particularly in the depression years.[9] And while the emphasis was on improving telephony, that mandate was defined broadly, although only a handful of scientists such as Davisson were ever given complete freedom to pursue "pure science," a term that, in any event, Jewett disliked. "The word 'pure' implies that all other kinds of research is 'impure,'" he wrote Bush in 1945.[10]

Jewett had been elected president of the National Academy of Sciences in 1939, the first engineer to attain that post, and during the war he joined Bush's inner circle at OSRD. Jewett shared a political point of view with his old friend and colleague Millikan, who, at the height of his fame in the 1920s, was an archconservative figure in American science—"the Nobel Babbit," in Kevles's words.[11] Jewett himself was a staunch Hooverite (when Bush and Jewett referred to "the Chief," they meant Hoover, not Roosevelt) who as early as the 1930s was arguing against federal aid to science, which, he claimed, would mean "a large measure of bureaucracy." In 1941 he gave a lecture at the University of Pennsylvania in which he claimed that the most common response to technical change, increased planning, was a threat to freedom bordering on dictatorship—a message very similar to the one in *The Road to Serfdom,* by yet another Viennese economist, Friedrich Hayek, which would be published in London in 1943. "Haphazard political methods must go," Jewett declared. "The operation of government on the basis of uninformed popular hunch and whim, coupled with political self-interest, can only end in absurdity, if not in disaster."[12] Throughout the war, he criticized the OSRD, which he viewed as an amateur effort that would upset the existing order. He also opposed the establishment of the Rad Lab, foreseeing future competition for AT&T. And at the NSF hearings, he was only one among ninety-nine witnesses to argue against any federal role in science. The first essential of "first-class fundamental scientific research, the only kind that is worthwhile," he testified, "was complete freedom for experimentation and operation unhampered by the limitations of a politically controlled agency."[13]

The examples set by massive R&D operations such as the Rad Lab, Bell Labs, and RCA Laboratories provided the best available justification of monopoly and mollified many New Deal liberals and economists. While most opposed big business and monopoly, most also accepted, on some visceral level, the notion that size and technological progress were allied, particularly as the cold war began and America found itself in conflict with the Soviet Union and its command economy. In 1951 a Harvard economist published a tract that attempted to describe how the American economy really worked. For an economics text, John Kenneth Galbraith's *American Capitalism: The Concept of Countervailing Power* proved extremely popular; it was a best-seller in 1952. Its author had been raised in an Ontario, Canada, farming village and had taught agricultural economics at the University of California before moving to Harvard in the late 1930s. There he collaborated with Henry Dennison, an industrialist and cofounder of the liberal Twentieth Century Fund, on a slim book urging more flexible antitrust policies, *Modern Competition and Business Policy.* After a heady stint as one of the key wartime pricing administrators, he spent the last few years of the war at *Fortune* (he was too tall for the military), before returning to Harvard in 1949. One of the first papers he wrote there was a fifteen-year survey for the *American Economic Review,* "Monopoly and the Concentration of Economic Power." That was followed by *The Theory of Price Control,* a book that received less attention than Galbraith thought it deserved. "I made up my mind that I would never place myself at the mercy of technical economists," he said later.[14]

Galbraith's distinctive prose style made him appear to be a breaker of idols. In retrospect, *American Capitalism* seems more a synthesis of current liberal notions, circa 1950, than a set of profound revelations about economic reality. Like Veblen, he emphasized the role of institutions and cultural norms over traditional neoclassical calculations of supply and demand; and like New Dealers such as Rexford Tugwell, he strongly believed in government planning. Galbraith cautiously accepted big business (though he heaped scorn on businessmen) in his theory of countervailing power—his economy of large, counterbalanced groups of buyers and sellers, workers and managers that so resembled the emerging cold war geopolitical blocs—sneering at the fears of antimonopolists and rationalizing the supervisory antitrust policies that had evolved in the late 1930s. Central to his scheme was his belief that innovation occurred only in large organizations. "Most of the cheap and simple inventions have, to put it bluntly, been made" he wrote confidently.[15] Like Schumpeter, though without his sense of dynamic change, Galbraith declared that large corporations alone had the financial resources to develop the kind of new products needed to power a modern economy. "Modern industry of a few large firms is an almost perfect instrument for inducing technical change," he wrote. "It is admirably equipped for financing technical development. Its organization provides strong incentives for undertaking development and for putting it into use. The competition of the competitive model, by contrast, almost completely precludes technical development."[16]

*American Capitalism* caused a stir. Galbraith was both widely praised and, particularly within the profession, sharply attacked. But for all the noise, only a few holdouts of the Brandeisian view criticized him for equating size with technological innovation. What most economists, including Galbraith, failed to recognize were centrifugal forces that were beginning to alter the traditional centripetal tendencies—toward monopoly or oligopoly—in a mature industrial economy. Capital was decentralizing, loosening the large corporation's stranglehold over Wall Street. And new technologies were undermining traditional industrial economics, particularly economies of scale.

## The Plight of Small Business

"Now is the time for all good men, regardless of party, to come to the aid of small business." Thus Rudolph Weissman beseechingly opens the first chapter, "Can Democracy Survive Big Business?," of his 1945 book, *Small Business and Venture Capital,* one of a flood of books and monographs that fed off the work of the TNEC. Weissman, a staff member of William O. Douglas's Securities and Exchange Commission (SEC), was no Galbraithian prose stylist, and his modest study was no best-seller. "The good society of the future which we seek in the face of war's terrible sacrifices cannot fulfill its part unless the necessary steps are taken to allow it the opportunity to function," Weissman droned on. "This means giving it access to capital, especially equity capital, the vigorous enforcement of antimonopoly laws, and encouragement in the way of tax relief or incentives." Weissman was also not as sanguine as Galbraith about a society dominated by jostling blocs of giant institutions and groups. "Either we wish to maintain free enterprise. . . or we desire to break abruptly with our traditions to embrace a society of big business only, with cartelization and monopoly as accepted features."[17]

Weissman had good reason to feel besieged. Nearly every survey available showed that small businesses were starved for working capital, dependent upon expensive, short-term financing, and suffering from high rates of mortality. Small businesses had few financing options: money raised from family, friends, or wealthy benefactors; bank loans; equity or debt raised in the capital markets. The cheapest of these— public equity—had surged during the giddy New Era markets of the 1920s, then all but collapsed with the crash. Now even Weissman admitted that there were serious obstacles to reopening the public financing window. Bond issues laid a heavy burden of interest payments on small and still-fragile companies; and equity issues could be sold only with the promise of a stream of hefty dividends, an outlay many small firms could ill afford. Although Keynes in his *General Theory* chastised the speculative fevers that he thought were part of an American character trait ("Americans are apt to be unduly interested in discovering what average opinion believes average opinion to be; and this national weakness finds its nemesis in the stock market"), most investors of the 1940s and 1950s were still focused primarily on longer-term bond or dividend yield, not on speculative capital gains. In other words, they were focused on payout, not growth.[18]

There were other factors that hurt small business. A clublike Wall Street, with its organized underwriting groups and fixed fees and commissions, demanded a heavy price for debt and equity issues. Weissman cited one study that showed that for every $100 of stock offered, $14, or 14 percent, went to SEC registration expenses, underwriting fees, and expenses—far above the prevailing Treasury bill, which was yielding less than 3 percent. (Wall Street firms tried to blame the high costs on the SEC; Weissman, however, sharply responded that registration costs amounted to only a few percentage points of the total.) That "price" varied inversely with the size of the firm and presumably with the risk. For companies with less than $5 million in assets, the cost rose to 17.9 percent; for firms below $1 million in assets, to 21.6 percent. "Such a formidable expense is a twofold obstacle [wrote Weissman]: first, reputable businessmen are reluctant to finance in this manner; second, intelligent purchasers of securities hesitate to invest in these issues."[19]

The outlook did not appear demonstrably better as the war ended. Wall Street was moribund. Initial public equity offerings were dampened by postdepression conservatism, by mounting taxes on higher incomes, and by the government's heavy financing demands. More and more money had begun to flow into large, conservative institutions, such as insurance companies, which avoided the stock market. Despite a rally in stocks in 1945, the market sagged again under the expectation of a postwar bust. Even large corporations generally avoided raising equity capital, opting for bond issues that they could then service from healthy postwar earnings; the price-to-earnings multiples of many blue chip stocks hung around four or five. The bearish dynamic was self-reinforcing. Corporate dividends were small, thus discouraging investors from buying equities, keeping trading volumes and share prices low, and discouraging corporate issuers from raising new equity capital. And what applied to large corporations applied doubly to small firms. The equity markets were simply unwilling, except at the most speculative fringes, to invest in newer ventures; and if one did succeed in raising equity down in the poorly regulated lower depths of the over-the-counter markets, the fees were often usurious, the liquidity elusive, and the crowd of local characters of uncertain moral standards.

Without a developed market in initial public offerings (IPOs), start-ups could either try to expand out of retained earnings (if there were any) or turn to the banks. Alas, the commercial banks, having tightened their credit standards after the relatively loose 1920s and the experience of massive bank failure in the 1930s, were not eager to finance flocks of precarious small businesses.[20] As they did in the early 1990s, commercial banks used the low interest rates of the depression to recapitalize themselves—not to lend. One wartime study of one thousand smaller corporations found that, as a result of such action, not only did most firms produce just enough profit to support their proprietors but the combination of economic slump and lack of credit had forced many of the survivors to sell off assets to cover losses.[21]

The weak IPO market also meant that venture investors had little opportunity to cash out. That was not only a disincentive for, say, wealthy investors, but it kept investment capital from circulating, thus limiting its beneficial effects. Indeed, the entrepreneur who lost his patron normally disappeared. Finally, there was no social reason either to invest in or to aspire to entrepreneurdom. The entrepreneur was not celebrated; indeed, he was considered a somewhat questionable character (as he is today in Japan) compared with the corporate executive.

And yet there were signs that the situation was changing. Various proposals—from the SEC, from Kilgore's early NSF bill, from Berle—floated the idea of direct aid to small or new business, perhaps by establishing regional credit banks, either attached to the Federal Reserve or capitalized independently. In May 1949, President Truman came forward supporting such a credit scheme. In a series in *Harper's*, Peter Drucker declared that a radical shift in the public policy away from "bigness" was occurring.[22] The president's proposals, Drucker wrote, "were certainly only the beginning of the attempt to replace the negative and punitive anti-bigness drives of the past with a positive and constructive small-business policy—an attempt which may well come to dominate American economic thinking for the next generation." It took longer than that, although Harvard, encouraged by Schumpeter (among others), did form the Research Center for Entrepreneurial Studies in 1948. Prof. Arthur Cole, an economic historian who had overseen the business school library since 1929, ran the center and enthusiastically attempted to interest other academics in the subject. While Cole did generate some interest, most economists continued to plug away at their Keynes; entrepreneurs remained a very secondary issue.[23]

More important than the talk of Washington was that the economy was awash in capital; consumers had been forced to save during the war.

The savings rate topped 44 percent in 1944, and by the war's end, total personal savings, less than $2 billion in 1939, reached $36 billion, nearly a quarter of all disposable income.[24] A report from the Minneapolis Federal Reserve in 1948 claimed that 73 percent of the companies studied were formed with the bounty of personal savings, and that conditions were not as bad as previous reports (meaning that of the TNEC) had indicated.[25] Pent-up consumer demand for durable goods such as appliances, cars, clothing, and housing also exploded after V-J Day. These newly dynamic sectors of the economy were only just developing in the 1930s and had lacked the power to pull the rest of the economy out of the slump; but now they began to expand at a great rate.[26] That boom was fed by the fifteen million GIs who returned from overseas service. Many of them poured into the universities as part of the Serviceman's Readjustment Act, or GI Bill. At a time when college enrollment totaled only 1.5 million students, 2.2 million GIs—some 40 percent of whom might not have attended college before the war—took advantage of the program. Although investors did not abandon the pessimism of the depression years as quickly as Galbraith, for one, wished, the kind of "spontaneous optimism" that Keynes claimed was essential to long-term investment in enterprise was stirring again.

## The Invention of Venture Capital

Ironically, perhaps the most significant new force liberating small business came from the patrician precincts of American society. John Hay "Jock" Whitney was the very paragon of a class, the WASP establishment, that had ridden the giant corporation to economic and social preeminence. Handsome, well educated (Groton, Yale, Oxford), athletic (*Time* put him on a 1933 cover astride his polo pony), he was raised to spend his money wisely. "Money has three purposes," a friend once reported him saying. "To be invested wisely, to do good with and to live well off of."[27] In 1929, his education complete and his father recently deceased, Whitney took a job as a $65-a-month "buzzer boy" at the brokerage firm of Lee, Higginson & Co. (turning down an offer from polo-playing friend Averill Harriman to train at Brown Brothers Harriman). He was perhaps the only clerk in history to come to work on a 65-foot yacht.

In the 1930s, Whitney invested his money in ventures that interested him, from Broadway plays (*The Gay Divorcee, Life with Father*) and Hollywood films (*Gone with the Wind*, through a stake in Selznick International) to corporations (Freeport Sulphur, Pan American World Airways,

Technicolor). He recognized his own tendency toward dilettantism. "I am always just a participant in things," he complained once. In 1942 he put his money into the hands of a professional money manager—Samuel Parks, a former banker with J. P. Morgan—and joined U.S. Army Air Force intelligence. In 1944 Colonel Whitney was captured by the Germans in southern France and held for eighteen days before escaping.[28]

Whitney returned home a hero. He was forty-two years old and richer than ever; Parks had successfully added to his fortune by investing in technologically advanced companies such as IBM (still a maker of electromechanical tabulating machines) and Minnesota Mining & Manufacturing. In 1946 he simultaneously set up two new organizations: the first a charitable foundation, the second a firm designed to invest money in new enterprises. From the start, the new-venture firm, J. H. Whitney & Co., with its elegant offices at Rockefeller Center, was a financial success. Whitney himself put up $5 million in seed capital, then added another $5 million a few months later. The firm's first investment was Spencer Chemical Co., an operation formed from a Pittsburg, Kansas, plant built by the government during the war to extract nitrates from natural gas for use in explosives. Kenneth Spencer, a former coal executive, wanted to retrofit the plant to make fertilizer. J. H. Whitney put up $1.125 million and Spencer Chemical prospered. In 1950 Whitney sold a portion of its holding for $6.5 million.[29]

Other early deals stemmed from some of Whitney's prewar investments. Some flew, others crashed. The firm, for instance, profitably assumed an earlier Whitney investment in Minute Maid, a company that had developed dehydration techniques to make frozen orange concentrate. Less successful was Sanoderm, a milk product developed by a dentist that was purported to cure acne. After trying unsuccessfully to apply dehydration techniques to Sanoderm, Whitney gave up.

What made Whitney's operation different, besides its sustained success, was his acknowledgment that the venture capital process needed to be professionalized. The amateur and the dilettante were out. J. H. Whitney began with five original partners (only Whitney put up capital and he received the bulk of the profits), including Whitney himself; Parks; financial adviser Richard Croft; former Lee, Higginson statistical department head J. T. Claiborne; and a Texas lawyer named Benno Schmidt, who would go on to become the firm's managing partner and a powerful figure in the biomedical research establishment. While Whitney played an active role in the early years, the rest of the staff, including a group of young associates recruited from MIT and Harvard, rigorously evaluated proposals that soon came pouring in. These experts (though they were far from "expert" in the early years) screened an average of five hundred proposals a year for the first decade and invested in sixty; thirty-one of them turned a profit. But Whitney felt strongly that the process really began only when the investment was made. Clients, he felt, should also be able to tap the firm's intellectual resources, to seek advice about questions of management and strategy.

Whitney's new enterprise, with its blend of idealism and profits, soon attracted imitators. His sister, Joan Payson, and banker Frederick Trask set up their own firm, Payson & Trask, often piggybacking on J.H. Whitney investments. And there were the Rockefellers. During the war, Laurance Rockefeller had donated enough money to keep the Rad Lab at MIT going while the military decided whether to support it. After the war, he convinced the family to set up an office that poured money into housing, electronics, and aviation (he backed McDonnell Douglas). Unlike Whitney's operation, the Rockefeller effort was aimed at more established companies and served as an adviser, not an investor, for the brothers.

If the venture business had evolved no further, it would have had little long-term economic significance. The sums that Whitney, Payson, and the Rockefellers invested were tiny compared with the total need. For venture capital to make any significant impact required a much larger cohort of financial intermediaries who could tap deeper pools of capital and channel them to appropriate new ventures.

Whitney had begun to professionalize venture capital. Another new operation, this one out of Boston, would take the first steps to institutionalize the business and give it its now-characteristic style. That company, ironically, took the same name as Raytheon's ill-starred predecessor, American Research and Development. The "new" investment firm of American Research & Development (ARD, as opposed to AMRAD) was set up in 1946 by four powerful, self-made Boston figures associated with major New England institutions: MIT president (and wartime colleague of Vannevar Bush) Karl Compton; Ralph Flanders, then the president of the Boston Federal Reserve Bank and later a U.S. senator from Vermont; Merrill Griswold, managing trustee of the Massachusetts Investment Trust, arguably the first mutual fund; and Donald David, dean of the Harvard Business School. Boston had seen earlier attempts to raise risk capital for local companies. In the early 1900s, the Boston Chamber of Commerce had raised a small pool of investment capital; and in the 1930s, progressive department store magnate Edward Filene (a cofounder of the Twentieth Century Fund with Dennison and a client and friend of Brandeis) led a group to form the New England Industrial Corporation.

Now these four were again searching for a way to encourage economic growth privately. The driving force behind ARD was Flanders, a mechanical engineer, the president of the Vermont-based Jones and Lamson Machine Company and a man who, by war's end, shared much the same Yankee worldview as Vannevar Bush. Early in the depression, the self-made Flanders, a prolific writer on a variety of engineering, social, economic, and political issues and much taken by Frederick Taylor's notions of scientific management, had been so shocked by the slump that he became a strong proponent of applying engineering techniques to social problems. The United States, Flanders wrote in 1932, "was approaching a new stage in human development—the self-conscious direction of the mechanism of economic and social life to ends of general well-being. The eye that has caught this vision is satisfied with no other." But with the coming of the New Deal, particularly after 1935, he drifted back to more traditional principles of American individualism and morality, which would motivate his most famous act: leading the Senate censure of Joseph McCarthy in 1954. In 1937 Flanders attacked national planning. In the 1938 election he supported Wendall Wilkie. In 1940 he warned about encroaching socialism. Flanders, a Republican, never became convinced that the New Deal economic solutions had "solved" the depression, though as late as 1936 he and Dennison were studying the new economic faith with the Harvard Keynesians; and he joined Dennison, Morris Leeds, and Edward Filene's brother Lincoln in a 1938 polemic, "Toward Full Employment," which was drafted by Galbraith.[30]

At the end of the war, Flanders was running not only Jones and Lamson but the Boston Fed. In his memoirs, he describes how his earlier service as president of the New England Council, a regional economic improvement committee, led him to recognize the confluence of two New England strengths: the concentration of world-class research facilities in the region and the presence of "the greatest accumulation of fiduciary funds to be found anywhere in the richest country in the world."[31] Why shouldn't that money be used to nurture research? In November 1945 he floated a proposal before a Chicago convention of the National Association of Securities Dealers that, contrary to the Investment Act of 1940, fiduciaries be free to put at least 5 percent of their funds in new ventures. "American business, American employment. . . cannot be indefinitely assured under free enterprise, unless there is a continuous birth of healthy infants in our business structure," he said. "We cannot depend safely for an indefinite time on the expansion of our old, big industries."[32]

Flanders, with Compton, Griswald, and David, set out to persuade institutions to contribute to the new fund. The group proposed to offer stock at $25 a share, hoping to raise $5 million in equity capital. Alas, a sagging stock market meant that they were able to pick up only $3.75 million (it took a contribution from Lessing Rosenwald, the controlling shareholder of Sears, Roebuck & Co., to put the effort over the $3 million minimum). But they established a principle, particularly by raising funds at local fiduciaries such as the life insurance companies, investment trusts, and universities. The practice of tapping institutional money would prove to be essential to the later expansion of venture capital. Over the next several years, ARD struggled to clear away state and federal regulation that forbade fiduciaries from investing in companies less than three years old. It was not easy; ARD was fighting the phantom of the depression at every turn. Not only were fiduciaries large, bureaucratic, and deeply conservative, but the regulatory and legal system emphasized protection and security of assets over yield. ARD started small, requesting only a small portion of institutional assets. And it argued that such investments were made in the public good.

To run the fund, the trio signed up Georges Doriot, a slender Harvard Business School professor with a tiny mustache who spoke in heavily French-accented English. The son of a French automotive engineer, Doriot served in the French artillery in World War I, graduated from the University of Paris, then came to the United States in 1921 intent on attending MIT, only to meet Harvard president A. Lawrence Lowell, who steered him to Harvard Business School. After several years at the investment firm of Kuhn, Loeb & Co. in New York, Doriot returned to the school as an assistant dean in 1926. He published his most scholarly work with another professor, Cecil Eaton Fraser, in 1932: *Analyzing Our Industries,* an intensive, analytical look at a number of American industries.[33] Doriot combined a strongly felt laissez-faire philosophy, common to the business school in the 1930s, with a feverish devotion to capitalism and the individual entrepreneur. In 1933 he gave the business school's commencement address (preceding a speech by Adolph Berle), lamenting the Roosevelt reforms but then typically berating his audience. "We face a change," he said, "and we deserve it. . . . Do we really want a change from Franklin Roosevelt to Huey Long?"[34] During the war, he served as director of military planning for the quartermaster general and then, significantly, as deputy director of R&D for the Defense Department general staff; for the rest of his life he was known as General Doriot.

Doriot was not a prolific scholar. He favored the pungent, cutting comment over the detailed study. He could be notoriously demanding: he once famously sent his secretary home because she wore a red dress he did not approve of. His favorite forum was an optional class of his own invention for second-year students, called Manufacturing. In fact, the class had little or nothing to do with manufacturing; instead, Doriot used it as his pulpit to shock his big-business-oriented students and to stir up an appreciation for hard thinking, hard work, and the big challenge. Over the years, an estimated sixty-seven hundred students sat through Manufacturing, including a large number of future venture capitalists. "The course he taught was ostensibly about how to run a company," wrote *Fortune* in 1967. "But in fact it consisted of a series of lectures expounding Doriot's views on life, business, and even on picking a wife. He hammered into his students' heads three main themes: self-improvement, attention to detail in daily work, and an active concern with the future."[35] As one student said, "It didn't matter what they called the course. It was all about Doriot."[36]

> He'd ask us to list the top ten companies we thought had the most outstanding futures ahead of them. Number one was General Motors. Number two was General Electric. Number three, I think, was U.S. Steel. Doriot read through the list, then turned to us with that disdainful French curl in his lower lip. "You deesghust me," he said. "Most of you don't even deserhve to be in my clahss. Yueh're not wuerth my time. Yueh're shortsighted. Yueh're status-conscouse. Yueh're borrrhing. You know something? I gave zat same question to a clahss fifteen years ago, and you know what the answers wehre? Number one was Zheneral Motors."[37]

Doriot framed his notion of the venture capitalist in psychological terms; at ARD, he saw himself as a sort of father figure to the immature companies he was funding. He was unabashedly a proselytizer. He argued that venture capitalists had to provide more than just money to new companies, that they had a responsibility to offer advice, direction, the kind of expertise that small companies do not possess and cannot afford to acquire. Like J. P. Morgan, Sr., he argued that what truly mattered in selecting his "Grade-A men" was not money or property but character. Like that other celebrated New England lecturer, Ralph Waldo Emerson, Doriot preached a kind of commercial transcendentalism.

Doriot popularized the notion of an entrepreneur as a unique, special individual with what might be called, in a more religious context, an inner light—a sort of justification by entrepreneurial faith. At ARD,

Doriot himself basked in the hagiography of entrepreneurdom. Late in Doriot's career, *Fortune* described him as if he were some incarnation of Max Weber's industrious Protestant. He was raised in a strong Lutheran family. He lived "simply" on Beacon Hill in Boston, rising at seven and working at home at night. He and his wife "adopted" the flock of entrepreneurs that he backed. Greed and its trappings were the signs of the devil. "Something of an ascetic himself, Doriot worries that his 'boys,' the young scientists and engineers, will succumb to the temptations of early success. He fears that their efforts will slacken and that they'll start buying twenty-cylinder Cadillacs, fifty-room mansions, go skiing in the summer, and swimming in winter."[38] On the eve of his retirement in 1972, he said, "Our business is to build up creative men and their companies, and capital gains are a reward, not a goal." (Doriot died in 1987 at the age of eighty-seven.)

In the end, the most important part of Doriot's message had less to do with making money than with the human qualities of the entrepreneur. Schumpeter had already defined those qualities: "To act with confidence beyond the range of familiar beacons," he wrote in *Capitalism, Socialism and Democracy*, "and to overcome that resistance requires aptitudes that are present in only a small fraction of the population and [they] define the entrepreneurial type as well as the entrepreneurial function."[39] For Georges Doriot, entrepreneurialism was far more common than Schumpeter believed, if often hidden. It could be found and taught; it was part of the democratic, free market experience.

## New Economies of Scale?

Venture capital in the early 1950s was but a mere blip on the economic screen. But other powerful trends were also silently unfolding as the cold war began. Bond market returns relative to common stocks peaked in late 1946. Capital began a long, historic shift into equity, particularly into smaller stocks, a phenomenon that, in retrospect, had begun to accelerate in 1943. By 1944, the dollar return on small stocks bought in 1926 passed that of common stocks; it would never fall back again in the postwar era.

And there were larger productivity trends unfolding—again very quietly. In late 1947, at the annual meeting of the American Economic Association, John M. Blair, an economist at the Federal Trade Commission who had earlier written several monographs for the TNEC, presented a provocative paper that asked, "Does large-scale enterprise result in lower costs?"[40] Blair admitted that large corporations seemed to be a permanent part of the American scene and cast doubt that "entrepreneurial decay" (a phrase from English economist Alfred Marshall) would erode their entrenched positions. "Is it possible," asked Blair, "to point to more than a handful of isolated examples of large American firms which have suffered a serious loss of position during the last thirty years?"

Blair could not give such examples, yet he went on to rattle off statistics and observations that appeared to indicate that the seemingly glacial corporate world might be changing. First, he pointed out that the scale of corporate operations had leveled off since 1919. Despite depression and war, despite the perception of increasing corporate concentration, average factory size had not appreciably grown, and it may even have shrunk if industries such as automobile bodies, rubber, chemicals, steel, and electrical machinery, with their enormous centralized plants, were factored out. This shrinkage was occurring particularly in industries that had arisen in the twentieth century, such as petroleum and its by-products, consumer merchandising, and machine tools. Blair, citing Lewis Mumford on the advantages of small plants, noted how large corporations were opting to build more and smaller facilities spread across America. Newer technologies, particularly those springing from the exploitation of electricity (electric motors, transformers, generators), seemed to be undermining traditional economies of scale that made the larger producer more efficient; such technologies were thus having a centrifugal, decentralizing effect.

> The decentralizing effects of electricity extend to the individual machine which can now be located wherever it can be most advantageously operated. This mobility of equipment is in striking contrast to the inflexible and centralizing effects of steam, which tended to crowd together as many machines as possible along great line shafts hung from the ceilings and carrying pulleys to which the individual machines were belted. With its new-found freedom, the individual machine can now work at its own rate of speed. . . . In short, the machine has tended to become independent of the size and character of the plant in which it happens to be located.[41]

Electricity was key to this decentralization, but there were other technological trends as well. Blair carefully delineated the advantages in ease of use and cost savings that new materials such as plastics, plywood, and light metals afforded over iron and steel. Multipurpose machine tools allowed firms to reduce the number of specialty workers and improve flexibility. New techniques such as welding, stamping, and die cutting reduced the need for extensive machining. Perhaps most important, industrial measuring, recording, and, particularly, controlling instruments (which were growing very fast) boosted precision and again reduced the need for specialty machines and their high-cost, specialized operators. The automobile and the truck allowed plants to be located far from rivers and railheads and removed from urban centers with their labor pools; workers could travel to and from the workplace by car. Although Blair was careful to plead ignorance of atomic power, he expressed the then-popular belief that dirt-cheap atomic power would give an unprecedented boost to these decentralizing, capital-saving tendencies.

If production was decentralizing, Blair then wondered, shouldn't decentralized ownership and control shortly follow? That was a leap. Although the economic studies were fairly old and incomplete, Blair found some evidence to suggest that if smaller producers were not dramatically more efficient than large ones, they were certainly not *less* efficient. Blair argued that, indeed, in cases where large firms had an edge—say, in baking—the edge could be traced to sheer size, not inherent efficiency. That is, huge volumes tend to reduce the man-hours per output, and marketing power tends to produce a monopolistic hold over retailers. Blair reiterated the case against large companies running many operations: it increased the difficulties of corporate management, reduced incentives for efficiency, widened the gap between labor and management, required expensive accounting and control mechanisms, and spawned inflexibility.

In short, Blair was predicting the downfall, or at least the erosion, of the hegemony of the giant corporation. These new technologies were altering the economics of production and reviving the forces of competition. Large might not be better—or permanent. If Blair was right, competition might be a hardier phenomenon than anyone suspected.

# 6

## The Entrepreneurial Transistor

*All electric forms whatsoever, have a decentralizing effect, cutting across the older mechanical pattern like a bagpipe across a symphony.*
—Marshall McLuhan, *Understanding Media*

### Anarchy and Insurrection

The modest, unprepossessing transistor may be the only technology of the postwar years to deserve the label of *revolutionary*. The transistor changed the face of electronics by supplanting the vacuum tube, then created a new industry based on its own, unique properties and dynamics. Its defining trait—a quantitative factor that produced qualitative transformations—was the sheer rapidity of its development. Even today, it is a remarkable phenomenon.

Beginning in 1948, the transistor evolved at an astonishing rate, not only in terms of new applications but in terms of capacity and power. While television, an electronic system, grew into a major industry, it did not trigger further changes throughout electronics; its applications, while undoubtedly rich in social effect, occurred vertically. The transistor, on the other hand, a component, led to integrated circuitry, then microprocessors, spawning the larger, extremely fertile field of microelectronics, which in turn invaded and transformed other technologies, industries, and economics. Solid state (of which semiconductors were just a part) remade television, as it would remake older electric devices, including the telephone and the radio. The computer took its own insurrectionary style from its semiconductor engine—setting up feedback loops of innovation. Without the exponential growth in semiconductor capacity, computerdom might have resembled the television industry, with its halting evolution and field of battle dominated by a few large producers.

The companies developing, making, and selling transistors exhibited those same dynamics. The new transistor companies swept past traditional vacuum tube makers. Television competition had occurred between substantial, self-sustaining companies that had built up financial war chests either in prewar radio or from wartime contracts; the transistor struggle took place among a melange of firms—small and large, new and old—a constantly shifting array more like the anarchic flow of a street riot than, as in television, the slugfest of a boxing match. These companies would feed off centrifugal tendencies described earlier—the decentralization of capital, the decentralizing economics that Blair described, McLuhan's "bagpipe across a symphony"—and new forces just emerging—notably, the great flood of government funding.

## Monopoly as Innovator

The entrepreneurial transistor was a product born of a monopoly, AT&T, and nurtured by a bureaucracy, the military.

In 1936 William Shockley, twenty-six, arrived at Bell Laboratories on West Street in lower Manhattan hoping to work on vacuum tubes under AT&T's Nobel laureate Clinton Davisson.[1] The telephone company had just installed the first crossbar exchanges in Brooklyn, a major advance in electromechanical telephone switching, and Mervin Kelly, the lab's research director, was already anticipating systems that would have to handle even greater loads. Although the new automatic switching systems allowed AT&T to reduce its army of switchboard operators, the demands of maintaining such a complex system, with its millions of vacuum tubes, was as problematic as earlier technological limits to long distance. Kelly suggested that Shockley ponder a design without mechanically moving parts, an all-electronic exchange. Shockley, whose interest at MIT had been the behavior of electrons in solids, soon fell into a collaboration with Walter Brattain, a member of the lab's solid-state group who was working with copper-oxide rectifiers. After several months attempting, and failing, to come up with a crystalline semiconductor device that could amplify like a tube, though without the tube's bulk, heat output, energy consumption, and limited life, the war intervened and much of West Street transferred to the company's New Jersey facilities to tackle radar. Shockley and Brattain both left to work on antisubmarine radar, though they were assigned to different laboratories.

Developments in the still obscure scientific field of semiconductors inched ahead during the war. In the 1930s, Walter Schottky at Siemens and Sir Neville Mott in Great Britain each had independently advanced semiconductor theory. And the Rad Lab for the first time brought together experimentalists, who had been trying to coax a response from various kinds of semiconductor materials, and theoreticians, who had been laboring to apply the larger framework of quantum mechanics to the field. The prevailing theory hypothesized that atoms of semiconductor materials had either an excess or a deficiency (a so-called hole) of electrons in their outer, valence rings. Under certain circumstances, that electron imbalance might produce a current such as in the presence of light (the photoelectric effect), a magnetic field, or an external electric current. Wartime research at Bell Labs had also established that this predilection for generating a positive or negative current could be coaxed along by implanting, or doping, specific impurities in a semiconductor. Particularly fruitful work came from a small group at Purdue University studying and purifying germanium. Soon after the Japanese surrender, Shockley and colleague Stanley Morgan paid a visit to Purdue's Seymour Benzer and Ralph Bray.[2] The differences between the outlooks of the two groups were significant. Shockley, who like Frank Jewett refused to recognize a distinction between applied and pure science, announced that his goal was to develop a solid-state amplifier, nothing more; academics Benzer and Bray expressed an interest in scientific insights into the nature of the semiconductor phenomenon that would stem from their germanium research.

Back at Murray Hill, Kelly named Shockley and Morgan coheads of the revived solid-state subdepartment. Brattain had returned, and Shockley convinced John Bardeen, a theoretical physicist who had served at the Naval Ordnance Laboratory, to come to Bell Labs instead of returning to academia. Bardeen, a mathematician by training who knew a lot about surface states in solids, was able to provide explanations for many of the results Brattain and Shockley produced. The three men made a roughly complementary team: Brattain the experimentalist, Bardeen the "pure" theoretician, and Shockley the theorist with an intuitive feel for what was going on at the quantum level.

Through 1947 the group groped along, helped by that dim torch of theory. Attempts to build the transistor Shockley had sketched out—a field effect transistor—failed, and Brattain and Bardeen went off to attempt a slightly different approach. In November 1947 ("the magic month," Shockley recalled later) the pair dipped a silicon crystal in salts and finally generated a small amplified current. Two days before Christmas the pair demonstrated a crude-looking device that resembled a germanium popsicle sprouting three wires. This device, the point-contact transistor, successfully amplified the current from one of the wires some forty times. AT&T kept the device under wraps for seven months, while patents were filed, manufacturing processes explored, and a name was chosen (*transistor,* for "transfer resistance," from Bell researcher J. R. Pierce). On June 30, 1948, less than a week after nervously telling the military of their discovery—the company feared the security-conscious military would want to keep it under wraps—AT&T announced the invention of the transistor. The world yawned, seemingly more interested in their new television sets. The *New York Times,* to its everlasting fame, buried the announcement on its radio page.

Excitement continued to run high at Bell Labs, and improvements came quickly.[3] By 1951 the lab had spent more than $1 million on the transistor. Shockley had dreamed up a more sophisticated junction transistor made of a single monocrystal (the earlier transistor was polycrystalline) in 1948, laid it out in a seminal book, *Electrons and Holes in Semiconductors,* in 1950, then supervised its development a year later. The field effect transistor, with its wires and electrodes, exploited surface effects and resembled the delicate cat's whisker of an old radio crystal. Shockley's junction transistor, in which the current flowed within the body of the semiconductor, pointed the way to modern solid-state electronics. Breakthroughs were also occurring in materials and manufacturing. Bell metallurgist Gordon Teal helped develop a system for pulling large monocrystals of silicon from the heated crucible, or melt, and then helped devise a way to carefully "dope" the crystal with impurities. Other developments contributed to the production of even purer silicon and more precisely controlled doping, both essential for the improved junction transistor.

## The New R&D Environment

Although born at Bell Labs, the transistor would soon be forced into a chaotic, rapidly changing commercial world. It was, of course, a dramatically different world from the 1930s—particularly because of the presence of government funding for R&D. In retrospect, it is impossible to imagine the subsequent evolution of the transistor industry without the massive new source of federal funding.

Government R&D funding, in all its many forms, nourished institutions desperate for capital in the decade or so after the war. By the early

1950s, the depression was just an unhappy memory; the economic warnings about maturity were fading, though they were apt to recur with each recession. After a five-year spending dip immediately after the war, federal money was again pumping into research laboratories and academia. Government expenditures for research and development fell from $2.6 billion in 1945 to $1.3 billion in 1946, then bottomed out at about $1 billion in 1947 (in 1953 dollars); that was still, of course, considerably more than the $216 million spent in 1940.[4] But with President Truman finally signing the NSF legislation, and both the cold war and the Korean War beginning, the money flowed again. By 1953 the government was pouring over $2 billion a year into research, out of total federal expenditures of $74 billion. The physical sciences, particularly physics, received the bulk of the money, 90 percent in 1953 (8 percent went to the life sciences, 2 percent to social sciences). Most of that, in turn, was funneled through two new agencies, the Department of Defense—which was backing new jet bombers, early warning radar systems, missiles, and submarines—and the Atomic Energy Commission, which was developing the hydrogen bomb and funding linear accelerators (the postwar successors to the cyclotron) and projects aimed at everything from nuclear airplanes to nuclear rockets.

Academia was particularly eager for its share of this largesse. Colleges and universities found themselves caught in a financial vise immediately after the war.[5] In 1947, the year of peak enrollment, over one million GI Bill students were on campuses; by 1950, that tide had receded, a falloff exaggerated by the wholesale flight of draft-age males as the Korean War began. The enrollment decline coincided with a deeper crisis in university finances. Costs had risen, but incomes had not. And a baby bust that had occurred in the early years of the depression was just reaching higher education; demographic studies showed that small enrollments might last as long as 1957. Tuition, which had accounted for about half of university income in 1940, now amounted to over 60 percent—making the enrollment slump potentially fatal for smaller, newer, or private institutions. Investment income had also fallen, from 6 percent to 4 percent, hurt by a lackluster bond market and by an inflation that had reduced the dollar's value by nearly half since 1940. The methods for overcoming these financial woes in many cases ranged from the draconian to the unlikely: reduce faculty; consolidate; seek out riskier, though potentially more lucrative, investments; raise more money; or lobby the government for subsidies.

R&D funding appeared as a godsend for universities that could get it. They rushed to build labs and recruit new faculty in order to attract new students and garner government-sponsored research. Science and engineering departments began a period of sustained growth. Spirited bidding contests began for scientific luminaries, particularly in physics (Princeton's Institute for Advanced Studies, the home of Einstein, snared Oppenheimer; the Rad Lab's du Bridge took over the California Institute of Technology). Universities scrambled to engage in "hot" projects; a linear accelerator building boom began. In engineering, students poured into the more "scientific" areas such as electrical or chemical engineering as the more empirical mechanical and civil engineering disciplines plateaued; the era of the shop-trained engineer ended. And some institutions took steps to build a corporate base around the university, like those that had grown up around MIT and Harvard, and to profit from spin-off businesses, such as real estate development.

At Stanford University, among the foothills and fruit orchards south of San Francisco, Frederick Terman, a professor of radio engineering and the son of the man who invented the Binet-Stanford IQ test, returned from the war with big plans.[6] Terman had grown up on the Stanford campus and had been the protege of the chairman of the electrical engineering department, Harris J. Ryan, a pioneer in high voltage transmission, and a business partner of MIT's Dugald Jackson. For graduate study, Terman headed off to Cambridge, where he studied with Jackson and his protege, Vannevar Bush, at MIT. Although MIT offered him a teaching position, Terman returned to Palo Alto in the late 1920s where he began preaching the merits of building up Stanford as a research center on par with the formidable MIT. Terman quietly encouraged his best students to return to the area to work and to set up companies. In 1938, Terman had arranged a $1,000 grant to lure a former student, David Packard, back from General Electric in Schenectady, New York. Then he prodded Packard and his old friend William Hewlett, who had trained at MIT before returning to Stanford, to begin making audio oscillators in Packard's Palo Alto garage for a company they named Hewlett-Packard. Terman both invested in the operation and joined the board.

With the coming of the war, Bush tapped Terman to head up the Radio Research Laboratory (RRL) at Harvard, a sibling of the Rad Lab whose mandate was to develop electronic warfare techniques for defeating enemy radar. Bush seems to have wished to balance the heavy physicist population at the Rad Lab by using Terman's extensive electrical engineering connections. Although the RRL was much smaller than the Rad Lab, Terman still had to oversee a staff of 800 at the peak in 1944; although a third had physics background, Terman mostly recruited from industry (a large contingent came from CBS's color television laboratory) and from academic electrical engineering programs (many from Stanford).

Back in California after the war, Terman, now dean of engineering, determined to build MIT-like links between academia and business. With his extensive ties throughout industry and the military—he was particularly close to the group organizing the Office of Naval Research after the war—Terman set out on what he called "steeple-building": creating a university reputation around high-visibility, well-funded, graduate research programs (Terman did not care all that much for undergraduate programs). In the end, Terman believed he would not only build up the university, but the regional economy. "The west has long dreamed of an indigenous industry of sufficient magnitude to balance its agricultural resources," he wrote. "The war advanced these hopes and brought to the west the beginning of a new era of industrialization. A strong and independent industry must, however, develop its own intellectual resources of science and technology, for industrial activity that depends upon imported brains and second-hand ideas cannot hope to be more than a vassal that pays tribute to its overlords, and is permanently condemned to an inferior competitive position."

In the early 1950s, Stanford's endowment had been battered by the depression, and it had few resources save for raw land around Palo Alto.[7] The university decided to develop six thousand of its eighty-five hundred acres by providing ninety-nine-year leases to companies such as Varian Associates (founded by former Terman students), Hewlett-Packard, and Eastman Kodak, thus slipping around a provision of its charter that banned the sale of land. Like big-time football, sponsored R&D was accepted as a new means of providing financial support for the university and the region. Like major athletic programs, science developed its own bureaucracy (at Stanford, Terman presided over it—part booster, part networker, part venture capitalist, part developer), its own appetites, and its own goals, which occasionally clashed with the traditional mission of the larger institution.

Government spending, particularly from the military, shaped the transistor industry in a variety of ways. First, the government directly subsidized certain technologies essential to national security. Electronics, increasingly ubiquitous in everything from missiles to radar to computing machines making calculations for bomb makers, was a prime recipient. Second, it directly generated freely available research in academia. Third, it produced a steadily expanding pool of academically trained scientists and engineers. In time, given the fluctuations of government funding and the economy, and tightness in the academic job market, many scientists would find the industrial lab and the executive suite to be welcome options to an academic career. Fourth, the rising tide of funding accelerated mobility by offering greater opportunities for researchers. In fields where science and technology were particularly close, cultural differences between academia and industry eroded: Terman was only one of the more visible academics to swing between commerce and the university. Fifth, while most of the money went to large companies (and elite research centers) some trickled down to smaller corporations. Indeed, that trickle-down meant more to small firms than the much larger amount meant to an already well-financed corporation. For large companies, federal funding supplemented internally generated capital; for small companies, that money—even in small amounts—was often essential to survival. And again, the military funding stream was particularly significant because small, new companies had been shut out of the public debt and equity markets in the mid-1950s and venture capital still lacked critical mass.

Unlike television, which remained a product of the industrial laboratory, the transistor blossomed in this new, complex environment. In 1951 the three military services assigned the Signal Corps to provide funds to improve the production and reliability of transistors. In 1955, the Defense Department spent $3.2 million on standard R&D contracts and $4 million to encourage transistor production enhancements.[8] The numbers, by later standards, were small, but the transistor industry itself was still quite small and costs were relatively low. The federal money fueled development, reduced the edge that large companies had long held over small firms (although, in gross terms, the larger, traditional electric giants received most of the money, even as they lost transistor market share), and contributed to the creation of a style and structure to the ensuing competition.

Government funding did not make transistor industry, but it did shape it. The transistor would have played a major role even if the new funding environment did not exist, but the structure of the industry might not have developed as it did. Like the railroad, the automobile, and the electric dynamo and motor, the transistor had an irreducible power. In time, it would have undoubtedly edged out the vacuum tube and remade the electronics business, but the pace of change decisively swung the advantage to the new, smaller, exclusively transistor-based companies. The rapid rate of transistor development eroded the economic advantages of capital, size, and reach. Particularly in the first, discrete transistor phase, it was relatively inexpensive to get into the business—although, of course, it was extremely easy to fail. But failures didn't matter in a world where the supply of new companies seemed to be inexhaustible.

Federal money, plus a small amount of venture capital, served as the vital accelerant. New and improved transistors, diodes, and rectifiers were developed quickly, filling niches and forcing firms to respond with unprecedented flexibility in research, manufacturing, and marketing. Ironically, automation came late to transistor production because it usually meant freezing the manufacturing process. Philco-Ford, for instance, developed its surface barrier and jet-spray production processes in the mid-1950s and confidently automated its assembly lines; by the time those lines were working, Philco had fallen behind. The rapidity of change forced firms to dabble with different management structures, decentralizing decision making and linking departments such as marketing, engineering, and finance in new ways. The demands of such rapid technical change required technical leadership; engineers and Ph.D.'s often dominated the new companies. Each product turnover created new opportunities to win or lose. This form of rapid-fire competition was quite different from the world described in *American Capitalism*.

The rate of change undermined traditional practices. Semiconductor patents had a limited utility; often they were awarded long after they had become obsolete. But even if a company was awarded a patent, it was often fatal to act, as Philco did, as if the underlying technology would hold for any length of time; the technology moved ahead too powerfully, washing away 20–30 percent of products every year with prices falling continually.[9] Those pressures increased when integrated circuits kicked in, miniaturization and capacity exploded, and development costs rose. Booms and busts occurred, not just as in television, every seven years or so, but almost monthly, across fragmented product lines that changed continually.

The economics of the industry was characterized by this obsession with speed and learning. Under the rigorous dispensation of the so-called learning curve, the first entrants in any niche normally possessed insuperable advantages because they would have more time to master the manufacture of a particular product—to generate mass volumes and economies, reduce rejects, boost yield—than firms that came later. Latecomers had to chase early arrivals on a gut-wrenching roller coaster ride of cost and price cutting. It was easier to think creatively about new products than to follow that tortuous and losing course. The future always beckoned.

## Transistor Diffusion

The pace of change was striking early on. Bell Labs' technical preeminence, like RCA's market dominance in television, eroded quickly, through no fault of its own. In the spring of 1952 the company held an eight-day symposium at Murray Hill on transistors for the electronics industry; thirty-nine companies attended. Still dogged by antitrust litigation, AT&T was not about to restrict access to the new technology. Besides, the more firms making transistors, the more royalty income AT&T would generate (AT&T charged companies $25,000 each to attend the meeting, an advance on royalties). The company's own complex patent ties spurred diffusion too. RCA and AT&T had extensively cross-licensed, and so RCA was soon licensing out the transistor on its own; a few smaller firms successfully gambled that antitrust problems would keep both AT&T and RCA at bay and took no licenses. Finally, AT&T gave away the technology for free to Raytheon for hearing aids—a gesture to Alexander Graham Bell's original interest in helping the deaf. By 1952 Raytheon was selling transistors to eighteen hearing-aid makers and had developed a tiny competitor, Germanium Products, in Jersey City, New Jersey, whose president expressed sentiments that combined prescience with hubris. "Trouble with the big companies is too many long-haired boys and not enough practical horse sense. . . . We expect to chase the vacuum tube price to hell and gone."[10] Not long afterward, Germanium Products merged with Bogue Electronics, a nearby Paterson electronics outfit. Neither would turn out to be major forces in the industry.

AT&T would continue to spread the transistor wealth. Throughout the 1950s and early 1960s, Bell Labs made a number of important breakthroughs, particularly in production techniques such as diffusion, planar, and epitaxial processing, which it either shared with licensees in symposia or simply passed along at regular Murray Hill meetings, which any licensee could request. That policy was formalized in the 1956 consent decree with the Justice Department, in which AT&T agreed to license all its patents, including those applying to semiconductors, for free. At that point, Bell Labs became, in effect, a public research laboratory (at least until the Bell System was broken up in 1981).

Bell Labs also provided talent to other companies. Although Shockley, Bardeen, and Brattain would share the 1956 Nobel Prize for physics, the commercialization of the transistor also involved metallurgists, physicists, chemists, and electrical engineers, many of whom found themselves in demand. Bardeen left in 1951 for the University of Illinois, anxious to pursue work on superconductors, an area the research managers at Bell Labs were not yet interested in; in 1972 he won a second Nobel Prize for his work in that field. Teal, who had pushed ahead with his crystal-growing experiments despite resistance from his bosses (including Shockley), decamped in 1953 for an obscure firm in Dallas called Texas Instruments, which had decided to try out the transistor business. And in 1955 Shockley himself resigned officially from Bell Labs to set up a transistor company in Palo Alto, where he had spent part of his youth.

By the mid-1950s, semiconductors were no longer deeply mysterious crystals. Semiconductor theory, constructed upon the by-now solid base of quantum mechanics, provided an armature, a context, for rapid development. Researchers were trying to make transistors smaller and faster, able to withstand greater temperatures, perform more and different functions; and having a common, empirically tested scientific model made an enormous difference. Production processes had moved ahead quickly. Baking on impurities ("diffusion") proved far more precise and efficient than implanting impurities in a crystal ("zone refining"). New product development quickly shifted from being a scientific endeavor aided by inspired engineering (or experimentation) to being an engineering project suffused with science (or theory).

The transistor, however, still faced a major obstacle to commercialization in the form of the robust, powerful vacuum tube industry. Unlike radio, which half expected to fall to television, the vacuum tube industry was confident of its staying power. The tube business was dominated by an oligopoly of firms known as the Big Eight, which included some of the largest, most powerful corporations in America: Raytheon, RCA, Philco, Westinghouse, Tung-Sol, CBS, General Electric, and Sylvania. Prices were fairly standard, and growth was steady as electronics expanded.[11] By 1950, tubes existed in a bewildering variety, in different sizes, and for a remarkable spectrum of specialized functions. The vacuum tube represented a large and sophisticated body of knowledge. Electrical engineers could design a tube for nearly any use by working with accepted rules of thumb, although making qualitative gains was proving more difficult and expensive. If some acute observers, like Kelly, saw that the limits to tubes were being reached, others were confident of further innovations. The spectacular success of television, dominated by a giant vacuum picture tube, suggested strongly that other, new uses for vacuum tube technology would be found and that healthy growth would continue. The major area of tube innovation was in miniaturization, which the military encouraged in two large projects in the 1950s.

The Big Eight did not ignore the transistor. By the mid-1950s all had entered the business in some fashion. Raytheon leapt in earliest, first with hearing aids, then with its extensive defense electronics work. By 1954 the remaining seven tube giants were also producing transistors. The field quickly grew crowded. A second group of companies with similar heft, though without vacuum tube experience, piled in: IBM,

Hughes Aircraft, Motorola, and Minneapolis-Honeywell. And then there were the newcomers, a swarm of obscure names such as Transistor Products, General Transistor, Radio Receptor, Transitron, Texas Instruments, Germanium Products.

Almost immediately, the transistor market demonstrated its fluidity. AT&T lost its exclusive hold over the technology. Raytheon was quickly supplanted by Hughes (revived by whiz-kidder Tex Thornton, later of Litton Industries) and then by two of the newcomers, Transitron and Texas Instruments. In terms of profitability, Transitron dominated the 1950s. But by the early 1960s, Transitron stumbled while Texas Instruments forged ahead, only to face new competition from companies such as Fairchild Semiconductor and then Intel, National Semiconductor, Advanced Micro Devices, and on and on. Few members of the Big Eight made much of an impact despite heavily outspending the newcomers on R&D, picking up the bulk of the government contracts, and generating over 60 percent of innovations. Between 1952 and 1968, the three largest generators of patents were tube manufacturers: AT&T (835 patents), RCA (668), and General Electric (580). But as emphasized by Ernest Braun and Stuart Macdonald, authors of the standard history of microelectronics, the smaller companies proved to be more aggressive marketers—they had no vacuum tube business to defend— and they were more effective in sniffing out and providing what the market needed and wanted. At nearly all the tube companies, the transistor business was run by executives who, at best, failed to recognize that transistors were more than just a new kind of tube, or, at worst, viewed them as a gimmick. A few of the new companies realized quickly that the transistor was a component around which a new kind of electronics industry would be built.

## The Triumphs of Texas Instruments

No one thrived as well, or as long, in this new competitive arena as did Texas Instruments.[12] The company, based in Dallas, had originally been formed after World War I as Geophysical Service, Inc., a small operation set up to explore for oil and gas with new techniques that bounced sound waves off geological structures. By the late 1930s, GSI had become one of the largest geophysical exploration companies around, although that wasn't saying a lot: in 1939, sales, which also included earnings from a small oil production unit, were small, and the company was still privately held. On December 6, 1941, the company completed a reorganization that consolidated ownership among a handful of top offi-cers, led by President Eugene McDermott and Vice President John Erik Jonsson, and talked of engaging in some diversification, perhaps in man-ufacturing. The next day, the Japanese struck at Pearl Harbor.

Hope quickly became a necessity. Oil companies immediately put exploration plans on hold, and sources of revenue dried up. Desperate, Jonsson suggested that the firm adapt its sophisticated exploration tech-niques to military needs, for signal transmission. McDermott traveled to Columbia University to work on submarine detection devices, while Jonsson went to the Bureau of Aeronautics in Washington to attempt to sell magnetic detectors for aircraft to the military. Soon, GSI picked up a few minor military contracts. The amount of money it made—about $1.1 million—was less important than the experience Jonsson and McDermott gained outside geophysical services. Both quickly recognized the postwar possibilities in military electronics; furthermore, Jonsson established a connection with a smart, young navy lieutenant named Pat Haggerty.

Patrick Eugene Haggerty was thirty-one years old in 1945, a native of North Dakota who had taken a degree in electrical engineering from Marquette University before becoming the assistant general manager of Milwaukee's Badger Carton Co. When the war began, he joined the navy's Bureau of Aeronautics and was assigned to the job of evaluating electronics companies, immersing himself in advanced technologies—an opportunity that would not have existed for him without the war. Jonsson met Haggerty in his search for military contracts, and the two men began lunching together. Haggerty, as *Fortune* later said, was "no research star," but he was extremely adept at working the frontier among research, industry, and government—what President Eisenhower would later call the military–industrial complex. He was a manager like Frank Jewett, not a scientist like William Shockley. He was smart, cool, highly professional, and a man who could impress those traits on an organization. In 1945, Jonsson began to talk to Haggerty about coming to Dallas to set up GSI's new manufacturing arm. Haggerty, mired in canceled wartime contracts (and presumably not eager to return to cartons), agreed, recruiting as well some of his Bureau of Aeronautics colleagues. In Dallas, he reorga-nized GSI's manufacturing operations and won a number of contracts for bombsights and upgraded magnetic detectors.

By 1951 Haggerty had doubled sales to $15 million, and the Korean War was boosting the company's backlog. By then, tiny GSI was under-going a metamorphosis. The geophysical business, while healthy, had declined in relative importance as Haggerty's manufacturing unit expanded. In 1951 McDermott, now chairman, and Jonsson, as president, decided to change the name of the company to General Instruments, only to discover, as Vannevar Bush had with American Appliance, that the name was already spoken for. After much arguing, they adopted the name Texas Instruments.

By 1951 Haggerty had begun talking about more diversification, preaching how Texas Instruments had to become "a big company," by which he meant sales over $200 million. In a lecture given in the early 1960s, Haggerty admitted that he had missed the invention of the transistor in 1948, though he would soon be acutely aware of what was going on.[13] By 1951 Texas Instruments was hounding Western Electric for licenses. When AT&T finally announced a licensing policy, Haggerty sent the $25,000 check a day later; in the spring of 1952, he headed north with a team to the first Bell transistor symposium. Back in Dallas, he set up a small semiconductor group that took over a small section of a Texas Instruments factory, and he began recruiting Teal, whom he had met at Murray Hill. Teal, a Dallas native, wanted to come home and run his own research outfit. Haggerty finally snared him, and Texas Instruments' real transistor efforts began.

Texas Instruments' research record was exceptional. In less than a decade, this obscure Texas company pioneered the first transistor radio and the first commercially feasible silicon transistor and shared the glory of developing one of the first integrated circuits, a semiconductor component containing more than one transistor or other device.

But nearly as noteworthy as its record of innovation is the disciplined way Texas Instruments balanced its financing with its developmental demands, marrying innovative technology to conservative financing.[14] It sold equity twice in the 1950s and took on a small amount of long-term debt and some preferred stock. It was a capital structure as well engineered, and efficient, as its transistors. The company tapped debt and equity markets only when it thought the time was right, not when it was running out of money; it used financial markets, it did not need them. It helped, of course, that its other businesses were able to provide funds for transistors (Texas Instruments never had to absorb a loss during its growth phase) and that innovations continued to appear. The government's purchasing program also provided important revenue, particularly when transistors were still more expensive than tubes. In 1954, over half of the company's sales came from a military that paid cost plus a nominal profit for transistors.

And, perhaps most significant of all, the scale of the transistor effort was still manageable. Transistor R&D and marketing costs never threatened to obliterate the company's resources. With only four thousand potential transistor buyers, the market could easily be covered by a fairly small marketing team, and research was much less expensive than, say, building complex electronic systems such as television sets. Texas Instruments spent about $1.15 million on transistor R&D between 1952 and 1955 and poured $3 million in plant and tools, assets "that would not have been salvageable," Haggerty later admitted, "if our strategy failed."[15] Such costs were a burden for a company with less than $1 million in profit, but "it was very modest compared to the impact that the strategy could have on the company if it were successful." The company was thus able to develop new products, market them quickly, generate the income to support further developmental work, and quickly move on. Equity offerings served as a sort of insurance policy, not as the first line of cash.

A close look at the orchestration of financing, of product development and spending, shows how Texas Instruments was able to hedge its financial risk. Equity in the company had been held by a small group of insiders, a perilously narrow foundation for a growing company. On October 1, 1953, coming off a year with sales topping $20.4 million (and profits of $909,000), the company sold 2.9 million shares at $5 a share and took a listing on the New York Stock Exchange. That offering came not long after it marketed its first transistor, a knockoff of Shockley's junction transistor, and around the same time it introduced a small radio containing its own germanium transistors. Thus was born the first "transistor" radio, the Regency, actually manufactured by the small Industrial Development Engineering Associates of Indianapolis. The Regency sold fairly well but never made much money for Texas Instruments (transistors still cost more than tubes, and the company had pumped in some $2 million in development and design costs); however, it generated credibility and provided experience with batch production of transistors. By this time, Texas Instruments was beginning to spend heavily to develop Teal's silicon transistor, a product most experts believed to be years away. Silicon was much more difficult to work with than germanium, but it could function at much higher temperatures, thus suggesting a myriad of new uses, particularly for the military. On May 10, 1954, Teal shocked a scientific meeting in Dayton by announcing that Texas Instruments had developed a commercial silicon transistor. He demonstrated it by playing a small phonograph containing a silicon transistor.

Success led to success. In late 1954 Texas Instruments was able to use its newfound credibility from both the Regency and its silicon transistor to sell preferred shares, which brought in a bit over $4 million (not including underwriting costs) and effectively covered the costs of the silicon transistor program; for TI, the price of that money resided in a dividend it had to pay investors. By that time, Haggerty knew that the silicon transistor would more than support the dividend. In fact, Teal's silicon work gave the company a near-monopoly position that lasted three years—an all but unimaginable lead by later industry standards and one that would have certainly attracted FTC attention had it been held by a larger company. By 1960, sales broke the $200-million mark, with the company racking up over 40 percent compound annual growth and holding a 20 percent share in the industry. It was now a "big" company.

In the early 1960s, Texas Instruments reprised its financial strategy to underwrite another fundamental breakthrough. While silicon transistors were taking off, the company had begun working on fabricating an integrated circuit. Research and production costs were rising, but so, too, were sales, particularly to the military. With the stock market bedazzled by transistor companies in 1959, TI was able to use some 640,000 common shares and a new issue of preferred stock convertible into equity to acquire Metals & Controls Corp., an Attleboro, Massachusetts, company that fabricated sophisticated metallic components for the electronic and nuclear industries. The timing was fortuitous. By 1960, Texas Instruments and the rest of the industry plunged into the same kind of overcapacity problem that periodically plagued television; the problem was also exacerbated by a cutback in military purchases during the early years of the Kennedy administration. While Texas Instruments and Fairchild Semiconductor, a new company from California that was its strongest rival in integrated circuits, struggled to get the new technology into production, companies found themselves stuck with rapidly devaluing inventory and depleted capital reserves. Profitability, which exceeded 7 percent in 1959, declined by 50 percent by 1963, and shares fell from a high of 256 in 1959 to 95 in 1961. Other companies began to bail out or fold up, and a consolidation began (the slump also contributed to Philco's problems and led to its sale to Ford).

Texas Instruments weathered the downturn, aided by its diverse product line, buttressed by its careful financing, and, finally, saved by its superior R&D capabilities. By the mid-1960s Texas Instruments was again climbing, powered by military and NASA purchases of integrated circuits as the space program expanded and the defense establishment raced to build ICBMs. By 1969 Texas Instruments was an $832-million-a-year company—the great bulk of that being earned in some form of microelectronics.

## Rise and Fall of Transitron

Not every newcomer made the right moves. The crisis of 1961 broke the steady ascension of Transitron, a Horatio Alger saga that was, for a time, more remarkable than Texas Instruments.[16] Transitron was founded in 1952 by two brothers, Leo and David Bakalar. When their father, a Lithuanian immigrant who taught school in Boston, died in the early 1930s, Leo, the eldest, dropped out of school and went to work for an uncle in the liquidation business ("a business," he later said, "that taught you to recognize value"). He soon went out on his own, then invested his profits in a factory that made plastics for shoes. By 1952 he had made his first $1 million. His younger brother, David, had studied physical metallurgy at MIT before taking a job at Bell Labs, where he found himself working on semiconductors. David was twenty-seven when the brothers decided to go into the transistor business. They did not bother to attend the first AT&T transistor symposium or take out a license, gambling that such patents might eventually be free.

The backing for Transitron came entirely from Leo Bakalar's plastics profits. Early development work took place in a corner of Leo's plastics factory; Leo and David interviewed early employees at Leo's Swampscott, Massachusetts, home. They hired their first employee, an engineer named Gunther Rudenberg, from nearby Raytheon. Within a few months, Leo bought and renovated an old knit-underwear mill in the town of Wakefield and a bakery in nearby Melrose, and Transitron was off and running. By the end of 1953, Leo, who claimed scientific ignorance, had poured about $450,000 of his own money into Transitron.

Transitron's greatest strength was in copying and improving, not inventing. The firm's big break came when it took a so-called gold-bonded diode made by AT&T—a slice of germanium with a tiny gold whisker, very similar to the point-contact transistor—and figured out how to manufacture it efficiently in volume. That breakthrough, in turn, led to the sale of large numbers of the diode to the navy, which installed them in shipboard fire control systems. By 1955, sales had grown five times, albeit from a small base of $1 million. By the end of 1956, Leo was able to pay himself back, keeping $125,000 in the company as equity. By 1959 Transitron raked in $6 million in profits from $42 million in sales, a 14.4 percent return on sales. Although Texas Instruments led in total market share and breadth of product line, the Bakalars boasted that they had a larger chunk of the so-called higher-priced military and commercial "quality" market.

Leo ran the financial side of Transitron, much as he had with his plastics operation. He bought the bakery and knitting mill cheaply, thus giving Transitron a low cost base. There were few secretaries, memos, or meetings. Transitron never built a marketing team, instead using its 150 engineers to sell a relatively short list of products. R&D costs were kept very low (in 1959, the military contributed $1 million for R&D, while Transitron spent $3 million of its own money). The company always assumed it could improve upon breakthroughs made elsewhere. It also made little effort to lock in its engineering talent, eschewing the kind of generous stock options some public companies were using. In 1959, it lost an important group of engineers to a start-up called Solid State Products, which then turned around a product Transitron had been working on. (One of Haggerty's greatest skills was in instilling a sense of loyalty to Texas Instruments, supplemented by stock options and high pay; the Bakalars never fully recognized how much talent mattered.)

Nonetheless, Transitron was the first semiconductor outfit that Wall Street fell seriously in love with. In August 1959 *Fortune* published a favorable article about the then-obscure company that speculated about a possible initial public offering. The anticipation mounted in the fall. The company opened talks with the NYSE, a coup for a seven-year-old company, and the SEC warned against manipulation of the intensely anticipated issue. *Business Week* noted how "brokers are calling the Transitron offer 'the hottest thing since Ford,' in reference to the Ford Foundation sale of 10.2 million shares [of Ford Motor] in 1956."[17] The Ford shares fell sharply back after the offering, the magazine noted, adding, "Whether the same will be true at Transitron remains to be seen." Finally, on December 11, Merrill Lynch Pierce Fenner & Smith, with one hundred underwriting associates, officially sold 13 percent of Transitron, or one million shares, at $36 a share. At that price, Transitron was worth $300 million on paper, nearly twice what *Fortune* had estimated in August, with a stock price some forty-two times 1959 earnings, well above blue chip stocks. The Bakalars had turned their $125,000 in equity into $34 million.

*Business Week* was right, however, about Transitron's future prospects. The public offering was a high point, although Transitron was able to sell another 1.2 million shares at $35 only ten months later. Then, military spending slumped, and the expensive, technically difficult shift to integrated circuitry began.[18] Throughout the 1960s, Transitron earnings and share price fell, though the Bakalars hung on, at one point trying to diversify and at another point, in 1978, loaning the operation $21 million of their own money. By the 1980s, Transitron finally abandoned the semiconductor business altogether, focusing on making a range of electrical cables, connectors, and other components. "We struggled with semiconductors and we actually tried to produce some of them overseas, but we had problems," said David Bakalar in 1983. "You've got to be at the very top of the technology, and even if you are, it doesn't mean you're making money. It was a long bitter struggle before we decided we were better off to put our money in other devices."[19]

## Shockley: The Rigors of Commerce

The Bakalars ran into trouble by treating the transistor as if it were a mature business. William Shockley, as famous as the Bakalars were obscure, failed by trying to reinvent the corporation. Shockley was undeniably brilliant, unquestionably difficult, with profound insights into physical processes yet a blindness to the people around him. He was not, like his colleague Bardeen, a world-class mathematician, and he often expressed his considerable insights in gnomic ways, which Bardeen and Brattain had to translate to the rest of his colleagues. His management style, autocracy disguised as meritocracy, was leavened by the structures and managers at Bell Labs. But once he went off to start his own company, he gave full rein to a confidence in his own essential genius that was never less than eccentric and, often, particularly in later life when he began to advocate his own brand of eugenics, obsessional.

Shockley had been born in London of American parents—his father was a mining engineer, his mother a mineral surveyor—who eventually moved to Palo Alto.[20] Thinking they could better educate their son at home, they kept him out of school until he was eight. Once he did enroll, he disliked it, gradually drifting through a number of educational institutions, including two years at the Palo Alto Military Academy and a short time at the Los Angeles Coaching School. He was clearly very bright. When he was in high school he and his family moved south to Los Angeles, where he became something of a physics whiz at Hollywood High. After a year at UCLA, he transferred to the California Institute of Technology, then took a teaching fellowship at MIT, and then went searching for Davisson at Bell Labs.

By the mid-1950s Shockley had grown restless. Although he was named director of Bell's Transistor Physics Research Group, he spent 1954 and 1955 as a visiting professor at Caltech. In 1954 he had a brief waltz with Raytheon, asking for $1 million over three years as a consultant; the deal went nowhere.[21] Soon he went off in another direction. He approached Arnold O. Beckman, the founder of a fast-growing scientific instrument company called Beckman Instruments, to see if he would

serve on the board of a new company he wanted to form. Instead, Beckman offered to back Shockley's operation. Thus Shockley Semiconductor Laboratories was set up, first in Palo Alto and then a few miles south,
in a small cinder-block building near the Mountain View Sears, Roebuck
& Co. Shockley had no problem hiring talent; his fame allowed him to
assemble what he called "my Ph.D. production line," a group, as later
events would prove, of enormous youth, energy, and talent. But almost
immediately, problems arose. Shockley clashed with his staff on questions of style and substance.[22] He tried to put into effect his own well-
meaning, if bizarre, ideas about how a commercial research operation
should be run. To banish secrecy, he posted everyone's salaries. In a sort
of extension of peer review to commerce he distributed weekly evalua-
tions and encouraged his engineers to evaluate each other regularly. And
he tended to personalize problems. He forced one employee to take a lie
detector test after making one technical delay (the employee passed).

More substantively, Shockley combined technical grace with commercial myopia. His consuming goal was to develop a two-terminal,
four-layer, subminiature germanium diode, an extremely fast switch that
could replace four separate components. The diode was sophisticated
technically but a commercial dead end. His staff urged Shockley instead
to focus on rapidly developing process technologies, particularly a
method recently invented at GE and Bell Labs called diffusion, and
emerging markets such as those for silicon transistors. Shockley refused
to budge. The situation deteriorated. "We kind of said, we work for a
Nobel Prize–winner, but we don't think he's very good," said one staff
member later. "We don't like him and he doesn't like us, but we like
each other."[23] And so in perhaps the most famous defection in U.S. business history, eight of Shockley's engineers—"the traitorous eight,"
Shockley called them—first quietly appealed to Beckman, who
remained loyal to Shockley, then abruptly quit to set up their own operation financed by Fairchild Camera & Instrument.

No neophyte to entrepreneurial challenges, Beckman soon discovered how treacherous this new business was. In 1957 he told shareholders in his annual report that Shockley was about to go into production
with his new diode. "The device," he wrote, "appears to have many
applications, and other devices are expected to follow in the near
future." In 1958 Beckman suddenly formed a new manufacturing company, Shockley Transistor Corp., and folded Shockley Semiconductor
Laboratories into it, absorbing a $3-million write-off and an annual loss
of nearly $1 million. "Semiconductor operations may reach the profit-
making stage during the current fiscal year," he wrote hopefully.[24]
Instead, in late 1959, Beckman sold the operation to Clevite, a Cleve-

land-based bearings, bushings, and electronics company that had acquired an early transistor start-up, Transistor Products, in 1953. Six years later, Clevite sold its transistor unit to Harold Geneen's ITT, which, in 1968, shuttered the Silicon Valley operation.

Shockley himself was undeterred by commercial setbacks. By 1958 he began lecturing again at Stanford, becoming a full-time professor in 1963 and reestablishing a consulting relationship with Bell Labs. As a commercial enterprise, Shockley Semiconductor had little impact. And yet, the demise of Shockley represents two realities of the developing industry. First, survival was difficult and prosperity was fragile. The sheer rate of change was so rapid that companies needed to make only one bad decision or so to fall fatally behind. Second, the Shockley experience demonstrated the limits of depending on military contracts. Although military funding helped accelerate transistor development, it also could consign its favorite companies to commercial and technical backwaters. The Defense Department simply would not—and could not—change as quickly as did commercial markets. By the time Shockley collapsed, transistors were becoming integrated circuits, and military funding was being replaced by the harder-driving civilian markets, where the action was. The smartest, most innovative, most greedy, most competitive souls drifted toward those markets.

Shockley's brief commercial effort also casts a light on another aspect of postwar technological economy: the emerging entrepreneurial ethos. Postwar America was extremely fluid, in terms of capital and people. Shockley's Nobel Prize was a magnet for funding and talent. He was able, because of his fame, to tap a drifting band of elite engineers more loyal to a particular technological imperative than to any large institution; the company died when those same engineers followed the imperative elsewhere. Shockley's Nobel also allowed him to flaunt a kind of management hubris, which combined an overweening self confidence with a near-total lack of management skills. Shockley failed miserably, but his attempt to combine business and academia, to manage R&D in a commercial environment, would return with a vengeance a decade or so later when dozens of new semiconductor and computer companies appeared, many of them declaring that they had discovered innovative ways of managing innovation.

With Shockley, the story of postwar technology begins to shift from Boston and New York, from Vannevar Bush and Frank Jewett, to Frederick Terman and the entrepreneurs of San Francisco and Palo Alto, from the structures and procedures of the large organization to the freewheeling enthusiasms of the financial markets and the entrepreneur.

# 7

## The Ethos of the Market

### Market Transit

The decade of the sixties was a crucial watershed in the postwar technological economy, for it represented the point where the financial markets developed into a major player in the process of setting R&D agendas. In the early 1950s, the opinion of the stock market would have
meant little among the multitude of decisions involving the placement
of R&D dollars; by the mid-1960s, that opinion had become much
more important, particularly in creating a lucrative environment for
venture capital and for financing emerging companies. After 1974 it
became the dominant factor not only among new companies but within
the mighty walls of the large corporation.

This shift has not been given its due. The new market style represented a palpable break with the old dispensation dominated by the large
corporation and the government: the presence of the financial markets
brought an increasing sense of urgency to R&D, a deepening volatility of
mood, a mounting emphasis on the start-up, on "revolutionary" technology, on the charismatic entrepreneur. The market also provided capital
that at first supplemented, then replaced funding from either a hard-
pressed government or corporate sources. The increasing importance of
the financial markets as a source of funding for new commercial technology also raised age-old fears about markets. Are markets all-knowing,
efficient processors of information, or are they dens of speculation,
wracked by fad and fashion? Were the choices markets made, by definition, correct, or was there a sharp difference between the interests of
investors and those of workers, executives, and the public at large?

The financial markets displayed preferences and prejudices daily and
openly. For example, by the 1960s, the equity markets were clearly a lot
less excited by the prospects for television than they were about the

transistor, though, at the time, the financial results of many transistor makers were not much better than that of Admiral or Philco. Television seemed to suffer from that terminal disease, maturity, which, according to the then-prevalent gospel of the Boston Consulting Group, made it a target for disinvestment, a "cash cow" whose profits would be channeled toward more promising investment opportunities by parents such as Ford or Rockwell. Wrote one investment magazine as early as 1957, only a decade after television's introduction:

> Television, though essentially a post-war phenomenon, is fast-moving into the grandfather stage. It would be an oversimplification to main-tain that for television there has been nothing new under the sun since its commercial development, but none of the advances made have been revolutionary in nature, and with the exception of the still untried color experiment, none has the capacity to stimulate the industry to repeat the performance that put 55 million sets in American homes in one decade. In that short period of time, the home set market has become virtually saturated, leaving the industry with an uncomfortable over-capacity. . . . The problem is simple but insurmountable.[1]

Investors believed in semiconductors, just as they, unlike the Japa-nese, grew to be deeply skeptical about television. Market prophecies, of course, have a self-fulfilling quality. Expectations of future growth, of continuing innovation—regularly satisfied by the semiconductor indus-try—produced a steady infusion of new hope, new money, new compa-nies, new products, new growth, and new expectations. Conversely, expectations of slow growth depressed shares of television makers, damp-ened investment, and eliminated the possibility of new, innovative, mar-ket-financed entrants. Ironically, while the market's optimism toward semiconductors helped fuel its own happy view of its growth prospects as an industry, it also helped spawn the kind of fierce competition that made it difficult for individual chip companies to maintain steady profits.

Wall Street often invests in companies based on their defining tech-nologies. "The most profitable single event you can hope to see as an investor is the birth of a genuinely great new basic industry," declared the promotional copy for a 1959 research report on electronics from the firm of Spears & Staff. "It happens only rarely (only five basic industries have been born in America since the turn of the century), but when it does happen, the money making opportunities are spectacular!"[2] Semi-conductor companies died regularly, and investors saw their stakes peri-odically go up in smoke, but the technological pace of change, the excitement, did not fade. When investors thought of semiconductors, they saw Texas Instruments, not Transitron; they saw Fairchild, not Shockley. With semiconductors invading new market niches, investors did not worry about saturation, as they did with television. Just as the pace of change continually spawned new opportunities for new products, so too did the proliferation of new companies create fresh, unblemished investment opportunities. Lastly, the rapidity of semiconductor evolution mirrored the accelerating turnover of the market and fed the drift of investors toward active investment management; like trackside bettors, investors in semiconductors could always look forward to a new race with new horses. And the obscurity of the technology (who on Wall Street really knew what was happening inside Texas Instruments, not to mention at newer, smaller, more obscure firms?) contributed to an increasing fixation on the dance of market prices rather than the close analysis of underlying operations, to what skeptical market observers from Keynes to Benjamin Graham defined as the difference between investment and speculation.

## Disciples of Growth

The stock market had been on a powerful upward swing throughout the 1950s, despite recessions in 1954 and 1957, reflecting economic expansion, fading depression fears, and a mounting confidence in the ability of Keynesian policies to master economic cyclicality. In 1949, with President Truman in office and a cold war looming, the Dow Jones Industrial Average, with its thirty blue chip stocks, had closed at 200. It ended 1959, with the Eisenhower administration winding down, at 679.4, a whopping 245 percent advance. It was the first leg of the greatest bull market in American economic history. More remarkably, Wall Street's view of debt and equity all but reversed. In 1950, dividend yields on common stocks averaged about two and a half times the yield on AAA-rated corporate bonds, reflecting the fear that stocks were dangerously risky.[3] By the end of the decade, dividend yields had fallen to only two-thirds that of bonds, a reversal of nearly four to one. What did that swing signify? "The emphasis had shifted radically from the risks associated with common stocks to their opportunities," wrote Benjamin Graham and David Dodd (with statistician Stanley Cottle) in the 1962 edition of their classic text, *Security Analysis*. "Investors were convinced that the same factors that produced the impressive growth in the 1950s would continue through the 1960s (and probably indefinitely), and that adequate action could be expected from the Federal government to prevent any downward spiral. . . . Furthermore, investment in most of one's funds in common stocks was regarded as virtually a necessity, to guard against the loss of real value through continued inflation."[4] Graham and Dodd were characteristically skeptical of this "average" view.

The equity markets were being fueled by a fashion for growth. *Growth* was a powerful catchword: the economy had been growing at over 4 percent a year since the war (at that rate, doubling every twenty to twenty-five years), with remarkably low unemployment, and the market had become entranced by capital gains growth (that is, profiting from rising stock prices) over dividend yield or payout. As a result, by the mid-1950s the stock market had begun to lavish its favors, in the form of sky-high valuations and multiples, on companies that had shown superior growth over a number of market or economic cycles. Although the exact definition of a "growth stock" was elusive—the number of companies, the percentage growth, the industries all varied—size and technological potential were two common benchmarks, particularly in the 1950s. Growth, both in corporate profits (or dividends) and in share prices (the two were tightly linked early on, though prices bobbed off on their own as the growth phenomenon took off), was another means of staying ahead of a boom economy's lurking shadow: inflation. As early as 1950, apostles of growth stocks, such as money manager T. Rowe Price, were urging investors to seek out companies with technological promise.

> The greater emphasis placed on research during and after the war has advanced our scientific knowledge at such a rapid rate that changes in the fortunes of whole industries as well as individual companies will take place more rapidly than ever before. Atomic energy will undoubtedly have an effect on some basic industries before the end of the next decade. Consequently many of the Growth stocks selected today will undoubtedly be replaced as new opportunities are afforded to investors.[5]

Price was wrong about atomic energy but right about the trend. By the late 1950s, growth dominated market thinking. In mid-1958, with the economy on the rebound, *Fortune* reported "a new kind of bull market" as institutional investors for the first time rushed into "quality" stocks, a euphemism for growth stocks. "Toward the end of May 1958," said one mutual fund money manager, "we awakened to the fact that the time had come to quit being conservative."[6] Investment in R&D-intensive stocks offered that hedge against inflation, but it also was instrumentality analogous to Keynesian economics: a strategy for mastering cyclicality, at least for selected companies. "In the absence of general business expansion, exceptional gains are likely to be made by companies supplying new products or processes," wrote Graham and Dodd. "These in turn are likely to emerge from research laboratories."[7] While sensibly warning that not every company performed research equally well, they took note of a study from the Stanford Research Institute investigating the relationship between corporate R&D numbers and return on assets. That SRI analysis of four hundred growth companies concluded that the relationship was positive and that "research and development expenditures have grown so rapidly because they are so profitable."

The first phase of the growth boom involved the market's affirmation that bigger was better; bigger was more technologically innovative, and safer. On nearly every list of growth stocks appeared traditional behemoths such as General Electric, Du Pont, General Motors, and Eastman Kodak or newer names such as IBM, Eli Lilly, Merck, Minnesota Mining & Manufacturing, Polaroid, and, as the 1960s rolled on, the fabulous Xerox (which introduced its first xerographic copier in 1959) and the conglomeratizing ITT, LTV, and Litton Industries. In nearly all cases, equity prices raced far ahead of dividend growth. IBM shares had advanced 1,216 percent in the 1950s, from 24 to 292, then in a remarkable run hit 579 at the end of 1961. At that price the market valued IBM shares at seventy-seven times 1961 earnings, more than three times the valuation of the Dow Jones Industrial Average.

The case for smaller technology stocks was more difficult to establish. To invest in smaller "growth" stocks required a higher level of market optimism. But by the end of the 1950s, confidence was rising—particularly in the products of the still-new venture capital business. A decade earlier, just as the postwar bear market was turning bullish, Georges Doriot had gone to Wall Street seeking additional capital for American Research & Development.[8] It was not a propitious moment. Doriot's portfolio of thirteen companies was still very young and still losing money, depleting the capital the firm had struggled to raise only two years earlier. Not only could ARD not pay dividends, but Doriot, in his astringent way, announced that it probably would not be able to do so for at least another four years. Despite a cover story in *Business Week* (the dapper Doriot stared from the page with the hint of a world-weary smile) and favorable notices in *Barron's* and *Newsweek,* Doriot failed to attract an underwriter for the offering and ended up selling only 43 percent of the proposed offering privately. It was not an encouraging experience for venture capital, a form of investment that needed the equity markets to provide capital for its incubating companies and liquidity for its venture investors.

A decade later, ARD boasted a portfolio of companies that looked considerably more robust—from Ionics, a firm developing membranes for desalination, to Airborne Instruments Laboratory, a Long Island defense electronics firm founded in part by Rad Lab veterans, to High Voltage Engineering Corp., a maker of electrostatic generators and one of the first venture capital–based technology companies to go public (in 1956). ARD had finally distributed profits to shareholders in 1954. Three years later, the company's board agreed to invest $70,000 for 70 percent of a new computer company in Maynard, Massachusetts, called Digital Equipment Corp. (DEC).[9] The company had been founded by two electrical engineers, Kenneth Olsen and Harlan Anderson, who had been working on Whirlwind and SAGE, two military-funded computer projects at MIT's Lincoln Laboratories. Olsen and Anderson had an idea for building a solid-state computer that was much smaller, cheaper, and easier to operate than IBM's giant mainframes, which, in Olsen's words, "were sealed behind two panes of glass"—that is, kept from direct interaction by users. Doriot first resisted the idea, but finally relented, reportedly to allow one of his younger staffers to learn his lesson (it was a small lesson: even then, $70,000 was not that much money). By 1959 Doriot's enthusiasm had increased. DEC, without having to tap more of ARD's capital, had made a small profit in its first year and was on its way to becoming what one venture capital observer would call "the investment that made the industry" and the one that certainly made ARD.

Although the focus here is on venture capital and financial markets, it is important to note how all sources of capital in the technological economy are interrelated. For venture capital to prosper, the equity markets must be prepared to step in and take over the burden of providing new capital through the process of going public. If financial markets are booming, venture capital can profitably cash out of its investments—and in turn can easily attract new investors. And if government R&D funding rises, private capital often follows to take advantage of newly created opportunities. ARD and the rest of the fledgling venture community (and by extension the equity markets) were thus buoyed by several key events that raised the overall level of technological funding in the late 1950s.

First, on October 4, 1957, came a small, silvery ball called *Sputnik I*. On November 3, the Soviet Union sent *Sputnik II* into orbit. The satellites were a profound, if groundless, shock to American technological confidence.[10] "Let us not pretend that Sputnik is anything but a defeat to the U.S.," wrote *Life* ominously. Handwringing about American schools, particularly science education, commenced, neither the first nor the last such bout but undoubtedly the most intense. At Senate hearings presided over by majority leader Lyndon Johnson, H-bomb developer Edward Teller testified to the need for more military technology; Jimmy Doolittle, head of the National Advisory Committee on Aeronautics, the predecessor to NASA, warned that the Soviets were about to forge ahead in many technological fields unless military R&D was increased; and an aging Vannevar Bush appeared to urge greater support for science, despite years of skepticism toward rocketry. *Sputnik* was "one of the finest things Russia has ever done for us," Bush said in his folksiest style. "It has waked this country up."

The *Sputnik* fears dovetailed with pressure from the science establishment for more funds for basic research. Post–Korean War military spending had leveled off, and Eisenhower, attempting to keep R&D spending under control by shifting basic research funding from the military to the NSF, had stoked fears that he was shortchanging academic science. Soon after the sputniks were launched, the administration boosted money for both basic science and science education and created the post of presidential assistant for science and technology. Moreover, Eisenhower approved a tripling of spending on big arms projects, particularly for missile and satellite development (both of which used miniaturized electronic components), from $2.2 billion in the third quarter of 1957 to $6.1 billion in the second quarter of 1958. The money served dual purposes: not only was it evidence that the government was "doing" something about *Sputnik,* but it provided a Keynesian jolt to a slowing economy. Less deliberately, it was a boon to both large and small companies, particularly in military electronics. The stock markets surged again.

The second event to encourage funding occurred nearly a year after *Sputnik I.* After decades of talking about helping small business, Congress finally passed the Small Business Investment Corporation (SBIC) Act in October 1958.[11] The immediate impetus was a Federal Reserve study that reported a shortage of equity capital for small businesses; the larger context, however, was a feeling, sharpened by the sputniks, that more private investment was needed to supplement the government's massive R&D efforts. The legislation that evolved, again heavily promoted by Johnson and his mentor, House Speaker Sam Rayburn, was compared at the time with legislation that spawned the National Recovery Administration and allowed specially organized companies to borrow from the government in order to invest in smaller businesses. An SBIC could raise funds by selling to the Small Business Administration, which oversaw the project, 5 percent debentures (the act mandated $300,000 minimum in paid-in capital, half of which could be raised by selling debt). Then, through further borrowing, the SBIC could leverage its capital up to four times ($1.2 million on that $300,000) and invest in small businesses by buying a debt issue that, at a future date, could be converted into equity.

Interest was higher than anyone expected, and money soon flooded into SBICs. Commercial banks, banned from lending for equity capital in the 1930s, were particularly eager to participate, attracting the wrath of Texas congressman Wright Patman, a dedicated foe of Wall Street and big banking and, ironically, a long-time supporter of legislation to help small business. Congress originally set aside $250 million, which, if fully leveraged, would mean at least $1 billion in investable capital. SBICs quickly dominated the venture business. In 1957, when Ken Olsen went looking for a venture capitalist, he could find only three—two in New York, one in Boston. By 1962 there were 585 SBICs, and counting.

There were several noteworthy aspects to the creation of the SBIC. First was the use of debt rather than the more conventional venture vehicle, equity capital. The debentures, on both the SBIC and the target company, were thought to instill discipline, although they also imposed a burden of repayment and an urgency to cash out that the patient Doriot eschewed. "Conservative" debentures were counterbalanced by the allowance for leverage, the concept of stretching capital by borrowing. Leverage had been associated with two businesses, both in varying degrees of disrepute: the 1920s practice of lending on margin to investors, which had been banned after the crash, and the banking practice of multiplying the effect of deposits by lending the same dollars out to many borrowers, a venerable technique that was viewed, particularly in the wake of the banking crisis of the 1930s, with some trepidation and tightly controlled. With the SBIC, it was agreed by Congress, the Fed, the administration, and both political parties (support was bipartisan) that some leverage in some circumstances was not dangerous at all—a quiet but decisive shift.

By the summer of 1960, DEC had just released its first real computer, the PDP-1, ARD was nicely profitable, and Doriot was clearly the vanguard of a new, glamorous, fast-growing area of finance: venture capital. It was now the conventional wisdom on Wall Street, in academia, in Washington, that a new age of technological development had dawned: it was "the research revolution," "postcivilization" (technological progress creating vast suburbs and a huge, happy middle class), "the third phase of the technological revolution, characterized by arrangements for systematically developing the art of management and by the rapid growth of specialized research." Technical knowledge, applied in an organized fashion, fueled a self-sustaining growth curve not just of selected companies but of the American economy at large; the fusty old limitations, from grim old Malthus to depression-era high unemployment to economic cyclicality itself, no longer needed to apply. The long-term planning required for effective R&D, plus "new products and processes, tend to cut across the ups and downs of business," argued one economist. "Technological research greatly increases the number of industries in an economy," wrote another hopefully, "[such that] no two have the same cyclical pattern." The result: smooth upward progress.[12]

This time, when Doriot, weighed down with technological credentials, called on Wall Street for financing, he was received differently. In August 1960, Lehman Brothers, one of the Street's elite "bulge bracket" firms, underwrote a public equity offering for ARD—the first for a venture capital firm—worth $8 million. It easily sold out in the early stages of a bubbling new-issues boom.

## The New-Issues Boom

The year 1960 marked the beginning of a sea change in market psychology. By the time ARD went public, the equity markets were gingerly, if excitedly, reveling in a new freedom. Certain lessons were now accepted as fact. First, stocks were not dangerous; actually, they were far more effective than bonds as hedges against inflation and far better earners over the long term. Second, science and technology would produce superior growth and progress. Third, certain industries, such as semiconductors and computers—electronics in general—were all but guaranteed to grow rapidly. ("Perhaps the most important aspect of electronics is that it is an open field with no apparent limit to the new function it can perform," wrote the *Magazine of Wall Street* in 1957. "We are now passing through an industrial revolution that dwarfs any similar period that we know of in recorded history.")[13] Fourth, blue chip growth stocks were becoming expensive, particularly for smaller investors.

Each of these "facts" pointed toward the attractiveness of the smaller, technologically based company. In 1962 an investment officer at First National City Bank named Arthur Merrill published a book called *Investing in the Scientific Revolution*.[14] Merrill's book was written not for Graham and Dodd's security analysts, nor for institutional money managers who were already fully invested in growth stocks, but for individual investors, who had begun to pour into the market in the 1950s. Conservative players such as Graham and Dodd, with their focus on uncovering tangible value beneath the shimmery dance of market prices, saw the benefits of R&D spending but were quick to warn the nonexpert away. "Broadly speaking, we think that modern technology has injected an important new factor in the affairs of many companies, which is not amenable to dependable prediction and which for that reason must be recognized as fundamentally speculative."[15] Merrill was more sanguine about the predictability of his "science stocks." He was careful to explain that while he called this "a new market," and thought it "a new epoch in stock market history," it was much safer than the infamous New Era of the 1920s. "This is not . . . another 1929 in which Radio Corporation of America went from $85 to $549 or Auburn Automobile from $78 to $514 on the basis of stock market speculation on slender margins against pyramided borrowings at high interest rates," he wrote. "International Business Machines, THE science stock of 1960 is owned almost entirely for cash in a virtually all-cash market. Not even the most cynical critic would suggest that IBM's stock could be pooled or rigged today, thanks to the Securities and Exchange Commission and the New York Stock Exchange."[16]

Merrill touched on most of the conventional themes. He noted that this new age was "not the industrial revolution but the Scientific Revolution. This is technological innovation organized *on a schedule* for a profit to create new products, new processes, new resources and new wealth"[emphasis added].[17] He was much more optimistic than Graham and Dodd about the ability of investors to master the dynamics of technological change. This new age demanded different analytical skills, a different economics; Merrill disparaged traditional investment benchmarks such as return on invested capital, dividend yield, book value, "bricks and mortar." He said he was not going to focus on standard corporate analysis of "car loadings, steel production or retail sales—or about Dow Theory," or any of the other schemes used to predict stock market movements. Instead,

> we are going to look at technology and try to develop a sufficient understanding of scientific fundamentals to appreciate what the technicians are telling us. We cannot possibly develop enough scientific knowledge to evaluate for ourselves any technology or black box, but we can develop a working knowledge of the language and basic principles involved so that we can be alert to important developments and quick to grasp what our technical consultants tell us.[18]

Merrill took the process one step further. He was not only interested in big technology growth stocks such as IBM, Litton, and Polaroid but was quite enthusiastic about small stocks, such as Kalvar, a spectacularly performing stock that was developing a process for quick photographic processing. He urged readers to master the tricks of recognizing scientific trends in order to get in early, to act as a venture capitalist like General Doriot, one of his heroes (Merrill printed a talk by Doriot and a description of ARD's portfolio in the appendix to his book). "New scientific discoveries are going on all the time, even in bear markets—embryonic Polaroids and new Littons are taking form somewhere. How are we to recognize them at a relatively early stage?"[19]

If anything, Merrill's book confirmed what the market had already discovered. A surge in new issues had begun to sweep Wall Street in 1960, climaxing in early 1961. In fiscal 1960, eighteen hundred new issues were filed with the SEC, double the number in 1958. More important, only a quarter of the 1958 total included issues for new companies; the rest were secondary financings. In 1960, however, one-half of the applications were for new companies, which required heavy SEC documentation, effectively swamping the agency in paperwork. That trend accelerated as the market picked up the pace in 1961. According to *Investment Dealer's Digest*, more corporate securities came to the market in the first half of 1961 than in any previous six-month period in U.S. history (although bonds amounted to two-thirds of the issues by dollar volume, as they had for some time). There was a particular emphasis on new technology issues, leading *Business Week* to dub the phenomenon the "sonics and tronics" boom.

*Fortune* was quick to pick up on the social implications of this phenomenon: the number of engineers and scientists who were suddenly accumulating vast paper wealth. "These egghead millionaires," wrote the magazine, are often founders or major shareholders of companies whose name "is likely to include such words as data, electronics, computer, instrument, control, space, systems, dynamics, research, microwave."[20] The magazine admitted that there were not many of them yet—perhaps a hundred (including David Bakalar, whom the magazine estimated was worth $150 million on paper)—but that their numbers were mounting daily. Perhaps more important, *Fortune* emphasized how new they were to the corporate scene, how unprecedented: it was the start of the construction of the New Entrepreneurial Man. "Never before have men so young, so unknown to the general public, and so insouciant about money been so eagerly embraced on Wall Street," wrote the magazine. "Most of these men are authentic eggheads; their values and preoccupations are still those of the scientific, rather than business, community. Some of them, indeed, are only foggily aware of the fact that they are millionaires."

The bull market in new issues had less fortunate implications as well. The flood of new issues, many bought by small investors, and the amount of money suddenly to be made selling and trading stocks, created fears of manipulation and speculative excess. New York Stock Exchange president G. Keith Funston, the man who had spearheaded Wall Street's "People's Capitalism" campaign in the mid-1950s to attract small investors, warned in May 1961 that "some would-be investors are attempting to purchase shares of companies whose names they cannot identify, whose products are unknown to them and whose future is at best highly uncertain." Funston was clearly irritated because many of the hot IPOs were showing up on the over-the-counter market, not on the NYSE or the American Stock Exchange and alerted by the Transitron offering, the SEC began a probe into the OTC markets. In late 1961 the market had crested at 735. By May 1962 it had lost 27 percent, declining to 536 in near-panic selling. The economy had slipped into a mild recession and the market had grown jittery about the Kennedy administration (the president had just been bullying Big Steel to lower prices, and there was talk of widespread antitrust actions).

Economists, fearful of the baby boom generation entering the work force and nervous about inflation, questioned the ability of the economy to continue both high growth and low unemployment. From 1947 to 1955 economic growth had cruised along at an average rate of 4.5 percent; from 1955 to 1962 that figure slipped to 3.5 percent. In 1961 a massive study by Harvard economist Simon Kuznets (the man who had led the government's first attempts to quantify the national accounts during the depression) historically quantified the relation of an increasing capital base to rising output and improved living standards.[21] Kuznets raised doubts that recent capital formation was up to snuff, that the historically unprecedented rate of growth could continue. Whatever the reason, investors moved to consolidate their gains and the market staggered. Typically, the biggest losers were the hottest growth stocks (IBM fell from 607 in December 1961 to 300 in June 1962), most spectacularly the highly speculative IPOs. "This period saw the complete debacle in a host of newly launched common stocks of small enterprises—the so-called hot issues—which had been offered to the public at ridiculously high prices and then had been pushed up by needless speculation to levels little short of insane," scolded Graham in the 1972 edition of his guide for individual investors, *The Intelligent Investor.* "Many of them lost 90 percent and more of the quotations in just a few months."[22]

A few months treading water did wonders. In 1963, with President Kennedy unveiling big military spending plans, the economy picked up again and the market, with happily renewed vigor, followed. For the moment at least, the qualms of Kuznets, Graham, and other pessimists could be discarded.

## The Allure of Technique

By the mid-1960s all the previous themes had resurfaced, though in exaggerated form: growth stocks, technology, and, later in the decade, initial public offerings. The go-go stock market embodied the most boisterous decade in finance since the 1920s.[23] Notions that once belonged only to the markets infiltrated into the corporate world, particularly a tendency to view corporate assets not as a legacy or as an organic collection of assets but as an investment portfolio—a basket of assets to be actively managed. The turmoil and change of the larger zeitgeist fed back into the market. Youth was a hot commodity on Wall Street, from young gunslinging mutual fund managers to "youth" stocks such as National Student Marketing. It was a market that gorged on "concepts" and "stories," particularly those involving science in any real or ersatz form. It was a market that viewed technology as a tool to shatter barriers, to disengage itself from the tug of an increasingly irrelevant past; it was, in the slang of the day, a "swinging" market. And it was a decade during which technological ideas and scientific terminology, particularly those suggesting that predictability was possible or that something could be created from nothing, were blended with marketplace practices and concepts, creating strange new hybrids.

Wall Street is always particularly attracted to concepts that, like leverage (a term borrowed from classical mechanics), suggest that money can be magically multiplied. By the mid-1960s, the conglomerate wave was already sweeping corporate America. Conglomeratizations, like the leveraged utilities of the 1920s, were creatures of the financial markets, though their architects rationalized them by extolling internal corporate dynamics that would release momentous energies. Most conglomerates were built on high stock prices and multiples; not surprisingly, the businesses that first attracted market favor were often technologically based, high price/earnings "growth" units—such as ITT, Ling-Temco-Voight, Litton, Teledyne, and RCA, all of which were started by exploiting the popularity of electronics—though later the glamour of the conglomerate concept itself was enough to keep them aloft. Acquisitions were made using stock; as long as the acquirer's multiple remained above the target firm, the game could proceed. But once growth slowed, once confidence fell, the market might abandon the stock, knock down its price and multiple, and force it to the sidelines. Not surprisingly, conglomerate builders such as Litton Industries' Tex Thornton and LTV's Jimmy Ling were brilliant manipulators of their company's images, masters of controlling both market sentiment and their own, at best, kaleidoscopic financials. As analyst John Wall wittily advised new conglomerateurs in *Barron's* in 1968:

> Get hold of the speeches and annual reports of the really savvy swingers who know the lingo and can make it sing. . . . You have to project the right image to analysts so they realize you're the new breed of entrepreneur. Talk about the synergy of the free form company and its interface with change and technology. Tell them you have a windowless room full of researchers . . . scrutinizing the future so your corporation will be opportunity-technology oriented. . . . Analysts and investors want conceptually oriented (as opposed to opportunist) conglomerates, preferably in high-technology areas. That's what they pay the high price-earnings ratios on, and life is a lot less sweaty with a high multiple.[24]

Like any pyramid scheme, the conglomerate game demanded growth at all costs; once mergers slowed, or stopped, the operational problems of most conglomerates surfaced. As a result, conglomerates were vast and growing constructions of disparate parts—car rental agencies, auto parts companies, steel mills, electronics firms, insurance units, military contractors, textile plants—held together by managerial concepts that appeared to be rooted in a modern, technical culture. If Raytheon's Laurence Marshall had envisioned a corporate assemblage linked by a common thread of technology, the conglomerateurs saw that assemblage woven together by a corps of "scientific" managers, trained in the latest financial techniques. There was little logic to the market's affections. On one hand it threw money at small companies run by scientists proudly lacking business training; on the other, it rewarded a concept based on nontechnical managers who boastfully claimed they could manage anything. Accounting, the language of finance, not electronics or another technical expertise, became the defining conglomerate discipline. Harold Geneen, the former Raytheon accountant who conglomeratized ITT, "treated money like a commodity; he turned ITT into a bank," one former executive said. "Geneen's product was money, and to that product he brought abiding loyalty," adds Robert Schoenberg in his biography of Geneen. "Since the sole object of his enterprise was to grow and to increase earnings, it made no difference what direction that growth took, whence those earnings came. Geneen was increasingly certain his management methods were valid for any business anywhere." Corporations, in this view, had no other constituencies but their capital-gains-seeking shareholders. As Schoenberg said, "Never a steel man or electronics man or telephone man, [Geneen was] always a pennies-per-share man."[25]

Advanced conglomerate theory explained that this new corporate form was really a Newtonian machine, a whirling constellation of counterbalancing businesses capable of diversifying cyclicality out of existence; conglomerateurs talked of building the perfect, frictionless mechanism that, like the idealized technological economy, ran by itself. The trait that made this work was, they explained, synergy, another word with scientific roots. Traditionally, *synergy* had simply meant a cooperative effort; it was adopted in the nineteenth century to refer to the orchestrated functioning of bodily organs and tissues. By the 1920s, physicians applied the word to the powerful combined effect of newly developed anesthetic agents; by the 1950s, it was used to describe the combined actions of two antibiotics or two anticancer therapies.[26]

The man who popularized synergy and gave it its full upbeat spin was Buckminster Fuller, the polymath and inventor of the geodesic dome and Dymaxion car who, despite a style of exposition that often defied comprehension, had gained a certain popularity as a social critic and philosopher in the early 1960s. Fuller, like so many others, believed that a new age had dawned—one of synergetics (synergy plus energy), first ushered in by Einstein, whose energy universe, Fuller claimed, overthrew Newton's universe dominated by inertia, and then further powered by self-conscious knowledge of how technical advances are made. Fuller defined synergy as the behavior of systems in ways that could not be predicted by the behavior of their subsystems.

By the mid-1960s synergy had become part of Wall Street's vocabulary. Tex Thornton reportedly picked up the concept from Beckman Instruments founder and Shockley backer, Arnold Beckman.[27] In the jargon of conglomerateurs like Thornton, it became known by the shorthand: one plus one equals three. Such mathematics survived only a few short years. By the late 1960s, the conglomerates had attracted antitrust attention and were beginning to collapse under their own weight and that of a sputtering economy. The outside world began to impinge on Wall Street's dreams. The market could live with the Vietnam War. What it feared was inflation set off by an administration intent on spending for both Vietnam and a Great Society. As interest rates rose, money tightened and merger activity declined. Although the bull market that began in 1949 peaked in February 1966, its last rally climaxed in December 1968, before the bears came out for good. In the last years of the bull market, IPOs staged a feverish repeat of the early 1960s rally. In 1968 and 1969 fifteen to twenty new issues a day went public, most of them getting an "instant premium," a sharp run-up in price that guaranteed short-term profits. But as the bear market began in 1970, the high fliers tumbled. On one gloomy day in 1970, young Ross Perot, the founder of Electronic Data Systems, a pioneering data processing company, lost over a billion dollars on paper.

Wall Street found itself mired in its own technological nightmare. While trading volumes and accounts had mushroomed during the bull market, securities firms had been slow to automate their back offices. Now with losses mounting, many firms found they lacked the capital to automate; a serious financial and paperwork crisis, followed by a dramatic consolidation, ensued. The old Wall Street club was coming apart. However, a new Wall Street would be built around a more powerful, more influential market than ever before.

## A Generational Shift

The early 1970s saw a more quiet, more personal transition as well. In 1972, ARD sold itself to Textron, a corporation based in Providence, Rhode Island. There was an irony here. Textron had been among the earliest conglomerates, assembled by its colorful chairman, Royal Little, in the early 1950s, though it had always been a bit stodgier—its original business was making rayon, hence the name Textron, although it later bought its way into military electronics—and more financially conservative than the high fliers. By the time the company bought ARD, Little had retired and the company was trying to convince analysts it wasn't a conglomerate at all but a "multi-industry" corporation.[28] Doriot himself was a still-vigorous seventy-two, but the ARD board had lost most of its founding members, such as Flanders and Griswold. As far back as 1957, ARD staffers told Olsen to inform the board when he presented his case that profits would come quickly, because so many directors would want to see results before they passed on. Times were undoubtedly bittersweet for surviving old-timers like Doriot. Although he had been successful, and lionized, some of his younger colleagues quietly criticized the old man for being soft, for hanging on too long to some failing companies and depending too heavily for ARD's return on the incomparable DEC (ARD's return on DEC was fully 8 percent of its compounded annual 14 percent return; the S&P returned 11 percent over that same period).

Such criticisms signaled a generational shift. As the go-go market rolled on, the giant institutions that once feared stocks of any sort were now enviously looking at the returns of successful venture operations such as ARD. Some of these institutions, of course, were simply corporations that saw venture investing as a way to track and tap innovation at small companies. Mostly, though, it was the glittering profits that beckoned. Like everyone else in the market short of the small band of Graham and Dodd acolytes, the institutions had gradually abandoned their fear of risk under the pressure to "perform, " to show competitive returns. And while venture capital was undoubtedly the riskiest of investments, its returns could also be the highest; DEC had returned 5,700 percent on ARD's original stake by 1970. And in a hot new-issues market, like that of the late 1960s, nearly anyone could drop some money in a new venture, then take it to the market and cash out at a hefty premium, the kind increasingly required to stay ahead of inflation. Although the SBICs had faded with increased regulation, big insurers (such as Massachusetts Mutual and the Travelers), diverse industrial corporations (such as General Electric, Du Pont, Singer, American Express, and the American Broadcasting Company), plus most major Wall Street firms had set up venture units by the end of the decade. Venture capital had also spread geographically, particularly to Palo Alto, California, where technological investors looked back East with the disdain of an ascendant class. "They don't vibrate in Boston like we do," said one Silicon Valley venture capitalist in 1970. "They're too establishment oriented, too structured, too concerned with dotting t's and dotting I's. Technology moves out here. Besides, how's the weather back east?"[29] Together, East or West, these institutions represented vast pools of money. Not only was the industry growing crowded and expensive, with more dollars chasing deals, but the style had subtly changed.

In 1970, *Institutional Investor* magazine, started a few years earlier to service the increasing numbers of large investors, contrasted the style of "traditional" venture capitalists with the new players. First, the traditional venture capitalist, like Doriot, hung onto many marginal companies long after it was clear that no significant economic return would develop. The reason: the industry was small and elite and relationships mattered. The venture capitalist had a responsibility not only to the entrepreneur and to his business but to other investors, most of whom he might have recruited. It was bad form to cut and run. Venture investing was, in Doriot's metaphor, a family. Competition mattered less than cooperation. Second, the traditionalist resisted the sirens of the market until the time was right—namely, when the company was strong enough not to be dominated by the ways of Wall Street. DEC did not go public for a decade, "to avoid the pressures of the public stock market," said Olsen.

The new venture capitalists, often investment bankers looking for new business as merger activity slowed or corporate officers with large amounts of capital to place, took what the magazine called the "one-in-ten" approach:

> The theory is that nine out of every ten venture investments will probably be losers, break-evens or mediocre gainers. But the tenth will hopefully be such a splendid winner that it will more than make up for the other nine. While these organizations investigate and keep track of their investments to a greater degree than they would a listed company, they do not see it as part of their job to assist the corporation in any but the most cursory way. Such assistance would be useless anyway—Value Line recently held positions in some 112 companies. If a company becomes sick, that is the company's problem, the normal, to-be-expected hazards of entrepreneuring.[30]

To the new crowd, with their large portfolios, responsibility ended when they could cash out in an IPO. And in fact, they pushed their companies to go public as early as possible, most before they had even turned a profit. Before federal regulations stepped in to halt the practice, many funds even sold pre-IPO "letter stock" to investors eager to bet that a successful IPO could be engineered. As the market for IPOs weakened in the mid-1970s, some venture money stopped going into early-stage deals at all and went instead into mezzanine financing, the last stage before a company is ready to go public; in effect, the funds were piggybacking on years of effort by others. Venture capital had become less investment and more deal making or, in Graham's phrase, speculation.

By the 1970s, market speculation had not only lost the censorious edge Graham applied to it, but it had been affirmed by academic theory. A new school of quantitative finance theory first appeared in the late 1950s, an outgrowth of the Chicago school of free market economics, and by the mid-1970s it provided a striking new way to look at investment and money management. This body of theory, with its capital asset pricing models, its betas and option-pricing formulas, rested on the notion of perfect, efficient, frictionless markets and assumed all the trappings of a hard science like physics. One of its more striking insights was the notion of risk as a function of a share's volatility against the market's volatility—the famous, and much-disputed, measure called beta. By building a diversified portfolio of different beta stocks, free market finance theory argued that investors could reduce so-called systemic risk much as conglomerateurs argued they could reduce risk through diversification.

Although first disdained as ivory tower nonsense by Wall Streeters (the academics argued that investors could never beat an "efficient" market over time, a theory that was quite a shock to the Street's self-esteem), the bear market of 1973, the worst since the crash, forced a general rethinking among investment professionals. They would not abandon the market. In fact, given raging inflation, fueled now by OPEC oil price rises, they needed capital gains more than ever. But they cried out for tools to control it; the new quantitative finance seemed to offer solidity in a treacherously shifting world. "Distress brought pressure for change throughout the world of finance: the way professionals managed their clients' capital, the structure of the financial system itself, the functioning of the markets, the range of investment choices available to savers, and the role of finance in the profitability and competitiveness of American companies," wrote Peter Bernstein in his history of these new tools in finance.[31] By the 1980s, the new finance, with market transactions at its core, was ubiquitous on Wall Street; it was orthodoxy in academia and common practice in corporations. Thus, the market not only determined which emerging companies were funded (and, by extension, which technologies received venture financing) but played a strongly influential role in determining which technological projects were funded by the larger corporations.

By then, the small, intimate world in which ARD was born had passed. Relationships were being replaced by transactions; the qualitative judgment of the strategic thinker was replaced by the quantitative numbers produced from market data. A financial world in gestation—on a restructured Wall Street, in venture capital, in corporations now looking at the markets with both fear and need—was larger, more complicated, more competitive, and more dominant. If government had stepped in front of big business just after the war, the markets had now elbowed past both as the key setter of the R&D agenda. By the 1980s the limitations of that new arrangement were becoming clear, most vividly in that hotbed of entrepreneurial market capitalism, Silicon Valley.

# 8

## The Balance of Microelectronic Power

### The Spirit of the Valley

In the history of Silicon Valley, Shockley Semiconductor Laboratory casts a far longer shadow than its short, uninspiring history suggests. Deeply unhappy, Shockley's dissident engineers had begun to think seriously about finding new jobs. One of the group was a manufacturing engineer Shockley had recruited from Western Electric named Eugene Kleiner, whose father had once worked for the New York securities firm of Hayden Stone & Co. That was qualification enough for the group to tap Kleiner to find a corporate home for them. The pitch letter Kleiner wrote landed on the desk of a Hayden Stone corporate finance executive named Arthur Rock, who had recently helped raise money for an early transistor firm, General Transistor.[1] In 1957, the same year Ken Olsen went seeking venture capital (and the year of *Sputnik*), the public equity markets for new companies were very thin, so Rock flew to California to talk the dissidents into seeking a corporate sponsor to set up a new company. One of Shockley's stars, Robert Noyce, a quiet twenty-nine-year-old MIT physicist who had previously worked for Philco-Ford, had not yet formally joined the dissidents, but he sat down and listened to Rock, thinking that the whole scheme seemed a bit far-fetched; he had figured that they would simply try to get hired as a group, not form a new company. Nonetheless, the group gave Rock a few weeks to go off and see what he could do.

After a dozen rejections, Rock came into contact with Sherman Fairchild, the scion of Fairchild Camera & Instrument and Fairchild Industries, two large companies based on Long Island. With a ten-page business plan, Rock, Noyce (who, once he agreed to leave Shockley, quickly assumed leadership of the group), and Kleiner then negotiated a deal with the company's president, John Carter. With $1.5 million from Fairchild, the dissidents told Shockley they were quitting, and they set up Fairchild Semiconductor in Mountain View. Noyce, a physicist, was first named head of research and development, then general manager. He impressed a scientific style on the new venture, though one quite different from that of Shockley. Fairchild, Noyce realized, was strictly a commercial venture. "Up to that point," Noyce later said, "it was very rare to take research people and motivate them by commercial success. Research people wore white smocks and were locked up in laboratories. Here the research people were out talking to customers."[2] Noyce's short time at Philco had also convinced him to shy away from military contracts, with their long development cycles, noncommercial specs, and fixed profits. "At the time I remember saying that selling R&D to the government was like taking your venture capital and putting it into a savings account— you're not going to make a substantial gain on it. Venturing is venturing; you want to take a risk and have the potential gains out of it," he said. "Our basic desire was to be out there venturing, and frankly, not to be beholden to other people you don't have that much trust in."[3]

Fairchild's founding partners had defected not only because of Shockley but because they were convinced that new manufacturing methods were the key to further transistor development. They quickly proved it. In 1961 Fairchild manufactured the first commercial integrated circuits—that is, a number of components, like transistors and diodes, embedded on a single silicon wafer. Jack Kilby of Texas Instruments holds the first patent on the integrated circuit, but like the first Bell transistor, it was a balky, demonstration device made of germanium, not a commercial product. Fairchild, in the person of Noyce, controlled the key manufacturing patents: a method of etching circuit lines on silicon called the planar process.

The development of integrated circuits solved a series of problems that had arisen from the mastery of transistor production in the 1950s. The transistor slump of the early 1960s had been exacerbated by economic recession and by the decline in military sales, which in 1960 amounted to half of industry sales; but it had been *caused* by production of discrete components that outstripped growing demand, squeezing profits and driving down prices. Integrated circuits would not eliminate boom-and-bust tendencies but would dramatically expand the market and accelerate innovation. Integrated circuits also answered the dilemma of systems designers, at computer companies and in the military, who faced the increasingly daunting task of wiring up many hundreds of thousands of discrete components, a complexity problem not all that different from Mervin Kelly's fears of a telephone system dependent upon millions of vacuum tubes. Integrated circuits provided ready-made blocs of already-wired circuits, particularly as more components were squeezed onto the same tiny turf.

Success came quickly at Fairchild. While Shockley was hemorrhaging, Fairchild Semiconductor posted $6 million in revenues in its first year, 1958; by 1965 it accounted for half the parent company's $180 million in sales, and nearly all the profits. With success, however, came the same problem that had felled Shockley. Fairchild began to come apart, albeit more slowly.[4] With only a 10 percent stake between them, the eight original partners had little to tie them to the company, and the parent did not see the wisdom of holding them or other employees. In 1959, six Fairchild employees left to form Rheem Semiconductor (acquired by Raytheon in 1961). In 1963 three of the original eight partners bolted to form a chip company called Amelco, soon to be bought by a growing local electronics firm called Teledyne, for which Rock had helped raise money. Other Fairchild engineers had a hand in companies such as Signetics, Molectro, Siliconix, and General Microelectronics (the latter acquired by Philco-Ford in 1966). Around the same time, Kleiner, Noyce's manufacturing chief, left to set up his own small company, Edex, to build teaching machines. In 1967 Charles Sporck, the man who had perfected the manufacturing systems for Noyce's integrated circuits, joined with venture capitalist Peter Sprague to revive National Semiconductor, taking Fairchild's marketing head, Donald Valentine, and a host of its manufacturing talent. The most devastating blow came in 1968 when Noyce, Gordon Moore (another Shockley refugee), and Andrew Grove left Fairchild to set up a new semiconductor company they called Intel. Fairchild Camera retaliated by spiriting away the entire senior management of Motorola's semiconductor unit, led by Lester Hogan; Motorola replaced Hogan with Wilfred Corrigan, who later replaced Hogan at Fairchild before going on to found LSI Logic. Soon afterward, W. J. "Jerry" Sanders III, a flamboyant Fairchild salesman, left to set up Advanced Micro Devices.

The famous diffusion of Fairchild talent tells only half the story. Fairchild was also the seedbed for Silicon Valley venture capital, which made possible much of what subsequently took place.[5] After arranging a number of other California deals, Rock moved to Palo Alto, forming a venture partnership with Tommy Davis, who had extensive California contacts, notably through Stanford's Terman. Rock and Davis backed Intel and, later, Apple Computer, among others. After working with Noyce at Fairchild and doing a stint with his own company, Eugene

Kleiner decided to try his hand at venture investing. He formed a partnership with Thomas Perkins, an MIT and Harvard Business School graduate and a Hewlett-Packard veteran (and a disciple of Doriot, who once offered him a job at ARD), to form Kleiner Perkins. Valentine left National Semiconductor to form Sequoia Capital, which led to two of the bigger deals of the 1970s: video-game maker Atari and Apple. Although many of the nascent valley venture capitalists, like Rock, Kleiner, and Perkins, had ties to the East, and to the Doriot gospel, the rootlessness that characterized the founding and dissolution of Shockley always lurked beneath the surface of both start-ups and established operations, accentuated by the money soon pouring into the valley.

Was it just California, with its gold-rush mentality, that created such volatility? Was it America, now hurtling through the affluent, disorienting, mobile 1960s? Was it the sheer youth of these engineers or the precocity of the technology? Or was it an amalgam of factors accentuated by the influence of the financial markets? Silicon Valley and microelectronics developed at the same time the financial markets were taking a larger role in the affairs of the technological economy, and they were intimately linked. The markets contributed to the instability, and creativity, of the valley. With their fascination for technologically based growth stocks, the equity markets poured money into electronic start-ups, peaking in 1968 and 1969 when Intel, AMD, and National Semiconductor were formed, then again in the 1980s with firms such as LSI Logic, Chips & Technologies, Cypress Semiconductor, and Mips. Venture capitalists, in turn, were much quicker to back new chip companies knowing that equity markets would be interested in them. Eventually, the equity markets saw the presence of certain venture capital firms such as Rock & Davis or Kleiner Perkins as validation, and another feedback loop was created. There is a correspondence between the dynamics of a trading-oriented, speculative market and the kind of dizzying job and corporate nameplate turnover that characterized the valley and the evolution of microelectronics itself. The deal, in both cases, loomed large. Fulfillment had to occur quickly; loyalties were fungible. Past experience meant little; what mattered was the trend. "Don't fight the tape," say the traders. "Don't fight the technology," was the mantra of the job-hopping valley engineer.

Is this hypermobility, variously dubbed "vulture capitalism" or "chronic entrepreneurialism," good or bad economically over the long term?[6] Silicon Valley itself split bitterly over this issue. Executives at established companies—including Noyce, Moore, Sporck, and Sanders, some of whom led famous defections—eventually bewailed such mobility, arguing that it damaged the industry; they even sued former employees for theft of intellectual property. Moore told shareholders that the formation of Sequel Computer in 1983 by a group of former Intel employees was "unmistakably unethical and wrong."[7] Firms such as Fairchild, AMD, and Intel, they argued, would have been much more competitive internationally if they had been able to retain engineering talent. Stability is necessary for long-term planning. Others, particularly from emerging companies, replied that the loss of talent more often than not represented a decline of corporate entrepreneurship and innovation and was a symptom of creeping bureaucratism. Mobility keeps change alive. Fairchild deserved to decline, they say, and the talent drain was just a sign. Defections may slow individual companies down, but they spur competition and technological innovation; they are manifestations of true, pure market competition, of Adam Smith's invisible hand. "What we need in the U.S.," said Cypress founder T. J. Rodgers in 1989, attacking the idea of semiconductor consortia, "are 14,000 semiconductor companies like the 14,000 software companies, working hard, going in a lot of different directions, competing like hell with some going out of business. But the industry will be so dynamic the Japanese won't be able to keep up."[8]

As Rodgers suggests, this argument is really one between the merits of large units and those of small, between a regulated "imperfect" market (financial, product, and labor) and "perfect" Brandeisian competition. It was eventually subsumed into the broader issues of industrial policy and managed trade that surfaced in the 1980s. If the government restricted overseas competition, encouraged consortia, or actively subsidized chip companies, who would it best help—established firms or newcomers? Would it enhance the start-up, and hence mobility, or bolster the defenses of the established operation? If it favored the small companies, would giant foreign competitors trample them? If it backed the large company, would innovation slow? In Silicon Valley, where change was most extreme and competition was most intense, the traditional divisions surfaced most clearly.[9]

The depth of those divisions points toward an industry no longer quite sure whether it was still in its youth or whether it had entered middle age. The gap between solid-state physics and solid-state electrical engineering had begun to close with the transistor, and the great commercial success of a physicist such as Noyce was just one sign that it had closed further; by the 1980s, the two fields were nearly indistinguishable. The unstated question of industrial policy, or of chronic entrepreneurialism, concerned the possibility of continued innovation, productivity, and progress on the scale that had characterized the industry in the past; it also concerned innovation that could undermine increasing cost barriers. To support industrial policy or condemn chronic entrepreneurialism was to take a pragmatic approach—that is, to protect the commercial monuments of the past and their current large employment and customer base. The belief that fundamental change would reorder the balance of power within the industry was a more optimistic view, one based on the notion that the most optimal scenario would be to junk and reassemble those monuments into new, more exciting configurations. In fact, as we shall see, the stubborn prosperity and hegemony of established chip makers strongly suggested that pragmatism was, at the very least, holding its own. The optimists were not about to be driven from this rapidly expanding field, with its pockets of fantastic growth, but they were not about to triumph either.

## Breaking Down Integration

Microelectronics in Silicon Valley advanced in a series of waves, each associated with a particular technological advance and a pattern of financing.[10] Fairchild begat a new generation of companies before fading from the scene. This new wave of Silicon Valley companies exploited integrated circuit design and manufacturing. The source of funds for new companies came predominantly from the military (as late as 1968, 40 percent of all output was bought by the military) or by corporate sponsors such as Fairchild Camera, Teledyne, and Raytheon.

By the late 1960s, venture capital had taken root in the valley. Noyce and Moore defected to Intel not only to escape interference from Fairchild Camera but to wrestle with the alluring possibilities of large-scale integration—that is, squeezing ever greater quantities of components onto a chip. Intel quickly pioneered the first memory chips and then the microprocessor, an entire computer on a single chip, and then took off. In its first year, the company had revenues of $2,670. By 1978 it racked up $399 million in revenues, and by 1988 that had risen to $2.9 billion. Between 1971 and 1980, Intel revenues and profits grew at a 65 percent compounded annual rate.

Intel successfully went public in 1971, an offering that was testimony to the company's strength, not to the robustness of the stock market. The excesses of the 1960s cast a pall over long-term investment, and economic turbulence dampened investor enthusiasm for long-term, illiquid stakes in new, small, and fragile companies. The IPO market had nearly disappeared by the time Intel appeared on Wall Street, and money going into venture investing fell below $50 million in 1975. The sale of General Doriot's ARD to Textron was significant and damning; the notion of a venture capital operation trying to operate as a public company appeared ludicrously impractical in the turbulent mid-1970s. The savage recession triggered by the 1974 oil shock left many overleveraged SBICs groaning under heavy burdens, a problem accentuated by mounting interest rates. The venture capital that did survive mostly formed private partnerships, funded predominantly by institutional investors.

By decade's end, a Democratic Congress and administration, intent on stimulating economic growth, scrambled to put into effect legislation designed to revive long-term investing. In the 1978 Revenue Act, capital gains rates were slashed from 49.5 percent to 28 percent; in 1981, under the aegis of the new Reagan administration, they were cut to 20 percent. Just as important was a series of reforms that went under the name of ERISA, which opened the doors for increased investment by those enormous pools of capital, the pension funds, expanding steadily as the population aged. In at least three different legislative initiatives in 1975, 1978, and 1980, the federal government defined what constituted fiduciary responsibility in public pension funds—and while those new rules limited pension fund managers in some ways, they also allowed at least a small percentage of assets to be invested in smaller, more speculative situations.[11]

But would conservative pension fund managers put their money into such inherently risky situations? In 1976, Roger Ibbotson, a University of Chicago finance professor, and Rex Sinquefield, a money manager at Chicago's American National Bank, published a study of financial returns in the *Journal of Business*.[12] Their conclusions: not only had stocks significantly outperformed bonds from 1926 to 1974, but they would probably continue to do so until the end of the century. Five years later, in 1981, the pair weighed in with another study that showed small-capitalization stocks outperforming large-cap companies over that same period. Those analyses, combined with the techniques of modern portfolio theory, provided pension managers with the rationale to begin to move some of their vast assets into more speculative areas, including venture capital and IPOs. By 1983 the pension funds pumped over a billion dollars into venture capital, more than five times the amount spent in 1981 and twice what the entire venture community had invested in 1979. In the peak year of 1987, all of venture capital would invest nearly $3.94 billion in some seventeen hundred companies.[13] Pension funding rose from 15 percent of the total in 1978 to 46 percent in 1984. The result, once again, was renewed pressure for bigger deals with a quicker payout. As the venture business expanded, the level of experience and skill fell, producing excessive valuations and a mania for popular concepts and deals.

The demand for capital grew, however, even as the supply increased. Intel had financed itself in the classical fashion: a dose of venture capital, rapid product sales producing retained earnings, then equity capital raised in a public offering, mostly to support new generations of products. By the early 1980s, despite the ready supply of capital, new semiconductor companies were forced to concoct more complex financial arrangements.

In short, new companies had to adjust to an era when capital expenditures were outstripping the ability of financial markets to fund development fully. In November 1980, Wilfred Corrigan left Fairchild to form LSI Logic, hoping to exploit new, application-specific integrated circuits, or ASICs, custom-designed chips made possible by the development of computer programs that automated much of the complex drudgery of integrated circuit design and by the invention of gate arrays, standard chips whose functions could be altered by turning circuits on or off.[14] Corrigan's timing was good: capital was plentiful, custom chips were fashionable, and he had all the requisite credentials. In early 1981, he easily raised almost $6 million from U.S. venture capitalists; a year later, he raised another $10 million from European venture investors. A year after that, in the midst of a thundering IPO bull market and just as LSI was recording its first profit, the company raised $138 million on the NASDAQ in a public offering. In the first three years of its life, LSI managed to amass over $154 million in capital.

Despite its rapid ascension to profitability, LSI's ongoing capital appetite remained voracious, not only because building wafer-fabrication plants and buying design tools was increasingly expensive but because of the inherently decentralized nature of its business. Manufacturers of standardized chips could build plants wherever they wanted: close to engineering talent or in areas with low taxes, cheap energy, and plentiful real estate. These units could be centralized, wringing out economies of scale and exploiting the learning curve with long production runs, and finished products could be cheaply transshipped overnight anywhere in the world. A custom chip maker was different. Like a supplier of bumpers or door handles to one of the auto companies, LSI Logic had to establish a close, working relationship with a limited, if dispersed, group of customers that had specialized needs and problems; to service those customers meant being near them, whether in Palo Alto, Osaka, or Frankfurt.

By necessity, LSI's financial structure was much more intricate than that of the classical chip companies—and most industrial companies. LSI set up subsidiaries in Europe, the United States, Canada, and Japan, each of which would tap local equity and debt markets for capital that would then be held for reinvestment by that subsidiary. Only the U.S. and Canadian subsidiaries were wholly owned, though LSI kept controlling stakes in all of its units.

What did such a structure offer to LSI? With capital requirements so steep, LSI had to look toward global, rather than national, markets. The emergence of global capital markets made financing on three continents feasible and reduced dependence on one market or one source of financing. Such a structure, however, limited its flexibility. The total capital of the company was always less than what appeared on its balance sheet because capital could not be easily shifted from one unit to the next.

After IPOs peaked in 1984, the financing environment grew harsher—and capital requirements still mounted. After 1987 and the market crash, new venture investing declined and new chip start-ups responded by extending the logic of LSI's financial strategy. Chips & Technologies, a custom chip start-up founded by Gordon Campbell, a former Intel engineer, went out and sought financial backing from potential customers, promising them future access to its products once they were actually manufactured. Once designs were completed, Chips farmed out manufacturing, reducing overhead and, in effect, stretching its available capital. Another start-up, Vitelic, pressed this partial integration strategy even further by working out a deal with Japanese carmaker Hyundai to provide capital in exchange for most-favored customer status. Vitelic designed the chip, had it made in Taiwan by a Korean manufacturer, and then considered an IPO in Taiwan.

Both companies reduced their overheads as early as possible, farming out every function except that which generated profits—the so-called value added. By running so leanly, smaller capital infusions, whether from a venture capitalist, a potential customer, or a strongbox beneath the bed, stretch that much further. "It costs at least $50 million a year to start a new computer company," said one industry observer in 1989. "That's a lot of money, particularly with Wall Street the way it is. Companies are thinking very hard about what they can do and what they shouldn't do. And they're getting rid of what they shouldn't."[15]

To many observers, this breakdown in integration appeared to be the logical, and happy, result of the technological and personal volatility of Silicon Valley. By the early 1990s, management theorists talked up the "virtual reality" company, in which all but the most essential functions are farmed out or shared in alliances, as another sign of advancing postindustrialism.[16] Perhaps, but the partially integrated company has its limitations and reflects weakness rather than strength. Specialization may be efficient, but it forces dependency upon a firm because it must share the rewards of innovation with many partners. Such dependency also spawns a complex strategic environment, with partners jockeying for position. In semiconductors, the partially integrated "virtual" companies of the late 1980s were implicitly conceding that they alone were not about to threaten the hegemony of the dominant companies. The founders of these companies might personally profit, since partial integration requires less up-front equity, but life in a partially integrated company is also more precarious and far more dependent on continual innovation than in a company with deep product lines and extensive ties to customers, distributors, and suppliers. The company that can integrate will do so. Those that cannot integrate have to live by their wits—never an easy proposition.

## The Establishment Ascendant

The waves of new-company formation, or "chronic entrepreneurialism," in Silicon Valley came to embody the Schumpeterian vision of creative destruction. A new decade appears, with new technology, new financing methods—and along come new companies, eager to slay their fathers. Each technological shift consigned marginal companies to oblivion, though stalwarts such as Texas Instruments survived; the eclipse of Transitron in the 1960s and the gutting of Fairchild in the 1970s and its decline in the 1980s were emblematic.

By the late 1980s, this Schumpeterian model still appeared to be functioning. Earlier in the decade, Japanese chip makers—Hitachi, Fujitsu, Toshiba, NEC, NMB, Mitsubishi, Matsushita—used their manufacturing prowess to dominate the market for memory chips, driving Intel (the inventor of the memory chip), Texas Instruments, AMD, and National Semiconductor out of the business and raising fears of another U.S. technological defeat.[17] Backed by enormous corporate agglomerations, funded by what American executives thought was unfairly cheap capital, and protected in their home markets, the Japanese openly talked of climbing from commodity memory chips to more profitable, more complex microprocessors. At the same time, the custom chip companies noisily announced that the age of the standardized chip company was over, that economies of scale had succumbed, and that the hope for American industry against foreign competition lay in embracing entrepreneurial customization. The established giants flinched, talking up research consortia and actively and successfully seeking trade protection. Intel tried to enter the custom business in the mid-1980s but, citing competitive pressures and its late entrance, retreated a few years later.[18]

In fact, the destruction of the large established U.S. firms did *not* occur—at least in part because of trade protection, but for a number of other reasons as well. Instead, by the early 1990s, Japanese chip makers found themselves absorbing huge losses, beset by overcapacity and a deficiency of capital and fallout from the collapse of the Japanese bubble economy and from the U.S. recession (they have not yet mounted a serious challenge in either the microprocessor or the custom chip business). Meanwhile, the custom chip companies have not inherited Silicon Valley either; they have either merged, restructured, or limped along. Intel, on the other hand, is so healthy and dominant that it has attracted Federal Trade Commission attention, concerned about possible antitrust violations.

How did this happen? Why did the cycle of creation and destruction *not* occur in the late 1980s? The reasons go deeply into the changing nature of microelectronics itself—and, of course, its relation to a larger financial world. The most apt metaphor is that of a community of nations. As in a global community, there are few rules to constrain competition—certainly none as restrictive as the television standards—and those that do exist are often suddenly washed away. Thus microelectronics companies, like nations, rise and fall as the environment changes, as their comparative advantage (in economic terms) either advances or declines. Relations are determined through perceived self interest; alliances are formed, and sundered; and the future, like that of some theoretically efficient market, is profoundly murky even for those entrepreneurial "visionaries" who find themselves on the right side of the zeitgeist. Winners today are often losers tomorrow. Change migrates, from sector to sector, exporting revolution, toppling long-established potentates, bringing in more and more companies to form an industry of increasing complexity and mounting strategic challenge. But as the whirlwind departs, it leaves behind an altered, if aging, set of companies.

How can we make sense of the forces that buffet this scene? Begin with the basic predisposition of the technology.[19] For over four decades, microelectronics demonstrated two broad, countervailing traits. First, semiconductor development was characterized by the exponential increase in capacity, memory, and processing power and an exponential decrease in price. The most famous description of this process is (Gordon) Moore's Law, which decreed that the number of components possible to put on any chip doubles every year. From 1960 to 1980 the maximum number of components squeezed onto a silicon wafer went from single digits to one hundred thousand. Intel's first memory chip, brought out in 1968, contained 1,024 bits of information; in 1993, companies were routinely selling memory chips containing over 16 million bits of data. In 1971 Intel's first microprocessor could crunch through information in 4-bit chunks; in 1993, companies were selling 64-bit chips. Exploding volumes also helped. Prices normally fell by 30–40 percent for every doubling of volume. This capacity growth tended to favor the small, fast-moving start-up with new technology.

At the same time, the industry saw capital requirements steadily increase as chip miniaturization and complexity rose. Texas Instruments spent perhaps $2 million to develop the silicon chip; by the late 1980s, Intel had spent $400 million just to build a state-of-the-art fabrication plant. Experimental X-ray lithography used to etch circuit lines less than a micron wide, much discussed in the late 1980s, required the use of a synchrotron that potentially would take costs into the billions (in the United States only IBM and the federal government could afford to build a synchrotron, although a number of Japanese companies could and did).[20] Despite ample venture and equity funding, these mounting costs favored established firms with long time horizons and deep financial resources; they forced start-ups to seek new ways of stretching their capital.

These two tendencies pulled the industry in opposite directions, though until the 1980s, the innovational drive far outweighed the countervailing growth in costs. The explosion in capacity and the decline in price fueled the rapid evolution of computing from mainframes (a business ruled so effectively by IBM through the 1970s that the government pursued antitrust charges) to minicomputers (DEC and Data General in the 1960s and 1970s), microcomputers (Apple, IBM, and many companies making IBM-compatible systems in the 1980s), desktop workstations (Sun Microsystems), and, in the last few years, laptops and handheld computers. By the late 1970s, this decentralization of computer power began to influence the underlying economics of semiconductors. Falling computer prices made automated design tools possible, which, in turn, reduced the cost of chip design, liberated the design function from the exclusive province of the large chip maker, and created opportunities for custom chip makers. Despite the design "revolution," however, semiconductor manufacturing costs still mounted, driving out marginal producers and giving an edge to the integrated, established firm.

This impressive wave of innovation tended to mask not only rising demands for capital, but also more subtly changing relationships between different kinds of companies and technologies in microelectronics. The primal division in the industry occurred between the systems house and the component maker. In the 1950s, most of the market power accrued to the systems houses, the giants of the old electrical machinery industry such as GE and RCA, computer makers such as IBM and Sperry Rand, and, of course, the military. Transistors were produced in batches as discrete devices and sold in bulk, like nails in a hardware store. The large systems houses then had to devise new ways to connect them into circuits and products. Early on, most opted simply to replace a tube with a transistor. But by the late 1960s, companies were designing hardware that exploited the advantages of the transistor over the tube. In this situation, the established, integrated electronics companies held a commanding position, not only because they had the financial resources but because they controlled market and distribution channels to consumers (or to the military); transistor makers were fragmented and prone to boom-and-bust pricing. It was thus easier to get into the discrete transistor business than to move into design and systems manufacturing; and it was simpler for a start-up to beat back RCA and GE in transistors than to drive RCA and GE out of the radio and television business.

With the development of the integrated circuit, the balance of power shifted. Circuits were now laid down on the silicon itself, forcing chip makers to develop *design* skills. As a result, some semiconductor companies edgily integrated into computer systems, mostly as a means of selling more chips, while a few computer companies integrated backwards into semiconductors in order to build better computers. Texas Instruments developed a small solid-state computer as a demonstration device and eventually sold radios, calculators, and computers. Fairchild, of course, built demo solid-state televisions, and Intel eventually sold everything from circuit boards to complete systems. Chip designers began to accumulate power. Even IBM, which had one of the largest chip operations in the world, had to build some of its computers from designs generated by outside chip makers. The precipitous drop in prices, plus the willingness of chip makers to take up the design burden, also provided an opportunity for smaller computer companies to break into the game.

In the late 1970s, the balance shifted again. The new design technology made it possible for any engineer with a computer and the right software to design chips. The number of chip designs mushroomed, and the custom start-ups appeared. By designing and building efficient, inexpensive chips or chip sets, the custom companies helped make it possible for small, obscure computer makers to cobble machines together that could underprice and outperform the largest competitor—even giant IBM. Observers such as George Gilder saw in all this design fervor the ultimate embodiment of entrepreneurial capitalism, a decentralizing trend that would erode the power of the large, established semiconductor companies and a revolution that would allow intelligence to triumph over capital.

> For the last five years [Gilder wrote in 1989], companies that aggressively pursued the Mead mandate for silicon compilers and similar tools have been gaining ground against firms that resisted the new regime. . . . Although some American chip firms protested mightily, the microcosmic insights of Carver Mead were transforming the industry in a way that deeply favored the United States. The new tools shifted the power toward systems design; the United States commanded nearly half the world's systems designers, four times the share of Japan. The devaluation of entrenched physical capital reduced the barriers to entrepreneurship.[21]

Gilder was correct about the centrality of systems design—and about American dominance—but he was incorrect about the eclipse of physical capital and of the coming transformation. Soon the battle over control spread to another important area of conflict: software. Chip makers first discovered that they had a stake in software when Intel invented the microprocessor in 1971, thus building not just blocks of circuits but the logic at the heart of the computer itself; that logic demanded a software program to make the microprocessor function. Although early microprocessors were fairly simple, and useful only for digital watches and calculators, the technology soon grew in complexity. While earlier products such as memory chips and integrated circuits were inherently compatible, microprocessors, with their software brains, were not. Now chip markets could be walled off into blocs and effectively conquered and controlled.

Unlike television, the computer industry depends on the market to set standards. The increasingly mass personal computing market quickly split into two blocs: an Apple world, with its Macintosh machines powered by a Motorola microprocessor, and a much larger IBM world (over 80 percent of the total market), driven by an Intel microprocessor and an operating software called DOS developed by Microsoft. Intel continued to introduce more powerful microprocessors, but its ability to control the marketing and distribution channels in personal computing increasingly meant just as much. By the late 1980s, both Microsoft and Intel had achieved the kind of market power in which success breeds success. Software writers and chip designers wanted to write for Intel and Microsoft because those two companies had the greatest opportunity for commercial success in the fastest-growing, most important computer markets. And, of course, commercial success was dependent upon getting enough software written for your "system" in order to attract customers.

Both Intel and Microsoft, ironically, were children of IBM.[22] In the late 1970s, IBM decided the time had come to exert its dominance over the gestating personal computer market, just as it had in mainframes, even though PCs were still a relatively small, not terribly profitable though fast-growing, business. As it often did, IBM established affiliations—this time with a software maker (Microsoft) and a vendor of chips (Intel) to provide the operating system and microprocessor for its first PC. IBM had many such affiliations; indeed, one of the fears of IBM dominance had always been its ability to destroy a smaller company by withholding or withdrawing its patronage. The IBM PC, introduced in 1981, was such a success that it established a market standard and propelled both Intel and Microsoft; but it was laughable at the time to think of either of them as potential IBM competitors. In fact, in the gloom of hard times in the semiconductor business, with Japanese competition mounting, IBM bought a 20 percent stake in Intel in 1982, publicly expressing the sentiment that IBM wanted to preserve a U.S. presence in semiconductors. Despite that investment, IBM forced Intel to share its microprocessor designs with other companies in order to ensure other sources of supply, a standard policy that also encouraged the diffusion of design and technology.

By 1987 Intel was openly defying IBM on a new generation of microprocessors, the 80286, by limiting the number of other manufacturers to which it licensed the designs. Intel then refused to license at all an even newer, more powerful chip, the 80386, which would drive the new line of IBM machines, particularly in its early, most profitable phase. Sales took off and Intel was already moving to commercialize the even *more* powerful 80486. Intel's ability to defy IBM on that second sourcing policy was a powerful sign (one of many) that IBM was losing control over personal computing, which in turn was rapidly undermining the old mainframe, minicomputer, and even supercomputer businesses.

By the late 1980s the computer market was still an IBM world, though one where IBM was more an embattled constitutional monarch than a despot. Although the IBM standard reigned supreme, the clone makers effectively limited its market share to less than 20 percent (there were some 350 IBM-clone makers worldwide in 1987, including a few, such as Compaq, that had grown quite large; all used Intel chips or knockoffs). IBM, in turn, tried to protect its position by ending its policy of "open architecture"—allowing anyone to design hardware options or write software for its PCs—and installing proprietary chips that outsiders could not easily copy.

But the game was already lost. IBM's attempt to wall off its system dampened the development of software from outside firms and reduced the popularity of IBM nameplate machines, which, in any case, were tied to rapidly fading technology. Meanwhile, other companies rushed into the vacuum, using the IBM architecture with more advanced chips and software. For Intel, the market outside IBM was now large enough that it could prosper without IBM sales; in truth, IBM needed Intel more than Intel needed IBM. Although IBM had built and sold machines with chips from Intel clone rivals and announced it would begin to sell IBM-made chips on the open markets, Intel now controlled the game.[23]

Microsoft underwent a similar and, if anything, more dramatic metamorphosis than did Intel. Microsoft's DOS had become a standard operating system for the IBM world.[24] But by the mid-1980s, the balance was shifting. In 1989 the alliance with IBM broke down (some observers speculate that Microsoft chairman Bill Gates deliberately scuttled it, leaving IBM committed to its balky, difficult-to-use software program called OS/2—developed by Microsoft for IBM—tied to an aging Intel 286 chip), and Microsoft poured its efforts into marketing its own software environment, called Windows.[25] Windows became such a dominant program that the Federal Trademark Office ruled in 1993 that "Windows" was a generic, not a trade, name. Because consumers flocked to Windows, Microsoft was able to induce smaller software companies to create programs that fit within the Windows framework, thus further ensuring its hegemony. And with capital costs rising on such complex standard-setting programs—Windows cost several billion dollars to develop—Microsoft gained an almost feudal domination over its core personal computer market. Such market control was reminiscent of that of IBM in the days when its 360 and 370 mainframes ruled the industry, and competitors quickly brought charges of anticompetitive behavior.

By 1990, the Federal Trade Commission, quiescent since the IBM and AT&T cases ended in 1981, began to investigate Microsoft; less than a year later, it also took off after Intel.[26] With $2.6 billion in revenues and 48 percent of the world software market, Microsoft cut prices ruthlessly and, critics charged, favored its own software programmers when it came to disseminating information about new programs. Intel, which controlled 68 percent of the microprocessor market, was charged with refusing to sell chips to computer manufacturers that also bought Intel knockoffs, or clones, and favoring its biggest and best customers at the risk of smaller personal computer makers.

Why didn't the decentralization of design tools open up the market instead of leaving it dominated by two behemoths? Because the fulcrum of control turned out to be not design but distribution, not unalloyed technological creativity but creativity plus marketing. The problem with decentralization is that it fragments; in a political dispute, with consumers as voters, fragmentation produces powerlessness. Like the lone inventor earlier in the century, the designer of a new chip architecture or the creator of a new software application lacks the power to extract the full value of his innovation. Such individuals, and such smaller companies, play the same role as that of lonely writers in a marketplace dominated by a few major publishers, a few large chains of bookstores, and millions of consumers voting with their wallets. Although individuals occasionally get rich, they will never get as rich as the companies that publish and distribute their work and control the channels to the mass public. In a maturing mass market, information age or not, the power belongs to the distributor.[27]

To seize and hold that kind of power requires a certain size and integration of functions: manufacturing, marketing, and R&D, not to mention a mastery of software and systems design. In their analysis of IBM's woes, authors Charles H. Ferguson and Charles R. Morris shift the metaphor a bit when they refer to the struggle for control over "architecture space" through the "knitting together" of a number of seemingly prosaic corporate functions. This architectural space is generally synonymous with the creation of standards, either market standards, as in computers, or regulatory standards, of the kind we see in television.

> The big winners in computers are companies that establish long-term, proprietary, general purpose, expandable, industry-standard architectures. Only a few companies have demonstrated, over an extended period, that they understand this principle—IBM in earlier days; Microsoft and Intel, the big winners of the 1980s; and perhaps a handful of others. Once a company establishes control over a specific architectural space, it can enjoy a highly profitable franchise whether or not its products are by themselves the best that the market can offer. . . . Its products, its manufacturing, its software must be very good, of course—that is the price of entry; but it is the strategic knitting together of these capabilities into a proprietary architectural franchise that is the consistent difference between the big winners and the also-rans.[28]

Given the complexity of the technology, this strategic struggle is no simple task. IBM was humbled because it made a series of strategic and management errors, not because of its large size or because economies of scale had suddenly been toppled; the tangled politics, infighting, and buck-passing of the large, dominant organization was but one factor of many. The technological world changed, and, like many companies, large and small, IBM misjudged that change. Its failure is just more obvious than usual.

In fact, the advantage of size persists. We already saw how capital costs were rising in both semiconductor development and software design and how an installed base attracts software support. But Ferguson and Morris point to a more fundamental trend rising from this seizure of architectural space: the tendency for computer markets to shift from specialized to more general functions.[29] As processing power becomes cheaper, the advantage goes to the system that can offer as many functions, or solutions, as possible, from word processing to graphics to games. Thus, Microsoft's Windows swept past the spreadsheet programs of Lotus, just as IBM in the 1960s seized control of the industry by providing a general-purpose machine, its 360. As complexity increases, as the integration of functions—and entire markets—continues apace, the barriers of entry rise. And yet because of the power of the technology and the eagerness of financial markets to fund new companies, the willingness to assault those barriers continues, giving the industry its characteristic battleground appearance.

What could threaten the hegemony of an Intel or a Microsoft? Various possibilities exist. The government could, of course, move to break the stranglehold of such companies, though that is increasingly unlikely, particularly under a Clinton administration eager to protect American high-technology exports. Alternatively, the government could simply distract and demoralize the two companies with years of litigation, as it did with IBM. The two companies could also turn on each other. Although they have often worked closely together, Intel and Microsoft could clash as each seeks greater control over the rapidly expanding world of personal computing. Microsoft is not likely to start making chips, but it could attempt to break Intel's hold by favoring competitors; it has already configured its most advanced version of Windows to run on rival microprocessors. And Intel has already moved to design its microprocessors to handle other operating systems that compete with Windows.

Continuing technological change could also lead to strategic or operational gaffes (like the kind that has felled IBM) or simply produce a decline in entrepreneurial drive. The continuing availability of venture capital, and of emerging markets and economies around a free market globe, will keep competitive pressures far higher than in earlier ages. As we saw with HDTV, the barriers among telecommunications, home entertainment, and computing are rapidly falling. As systems migrate to more general uses, it becomes difficult to tell which functions will provide the fulcrum of control; each stage in turn involves a fierce new struggle. Microsoft is working hard today to use its powerful base in personal computers to strike out and establish forward bases in networking, into database management, personal communications, and software development. Similarly, Intel must continue to develop new advanced microprocessor "engines" that also can operate on any of the warring operating systems, and it must attempt to stay ahead of the clone makers, including the Japanese and major competitors.

Both companies may fall, but even if they do it will not represent an indictment of size. Economies of scale have not been overthrown, except on the margin, and in mass markets such as personal computers the gap between science and technology has effectively closed. What produces uncertainty and threatens hegemony is the rapid rearrangement spawned by a technology moving through different markets, each creating ample opportunity for missteps, and the willingness of financial markets to finance new competition. Take either one of these two important factors away and you are left with a landscape more like that of the 1950s (or the 1930s) than of Silicon Valley in the 1990s.

## The Self-fulfilling Dream

The forces unleashed by microelectronics continue to sweep across the landscape, invading old markets and creating new ones. In each of these interlinked markets, the cyclical dynamic repeats itself. Growth, change, and rearrangement ripple through; in the confusion, small companies gain opportunities and the markets become the only viable setters of new-product agendas. But as innovation slows—if only from inevitable growth and the exhaustion of easy and straightforward technological possibilities—costs rise, at least until more fundamental technology arrives to shake things up again. Over time, new opportunities will decline and nonmarket financial power will reassert itself, just as with the major technologically based industries of the Second Industrial Revolution. It is too simple to pronounce the inevitable triumph of the small, the new, the entrepreneurial. After all, for all the recent difficulties faced by IBM and DEC, several venerable corporations (such as General Electric and AT&T in America; Siemens and Philips in Europe; and Hitachi, Matsushita, Toshiba, and NEC in Japan) have demonstrated an ability to change, compete, and survive. AT&T in particular has even been able to transform itself (with a little help from the trust busters), to threaten domination of a radically changed world created by the merging of computers and semiconductors with telecommunications and television.

Just as the potentiality of the chip created this whirlwind of change, limitations on further chip development will someday restrict it. There are two general kinds of limitations. First, there are physical barriers. How much can we continue to squeeze onto a thin wafer before encountering serious quantum difficulties? How much complexity can we handle, in both hardware and software? Predicting these boundaries is a fool's game, and we are likelier to come up against the second type of limit—financial—long before we hit the physical one. Companies, large and small, are already feeling the financial pinch of mounting development budgets. Costs will undoubtedly continue to mount, unless development ceases entirely. The ability of companies to innovate will increasingly depend on whether they can tap large amounts of cheap capital, although there will undoubtedly always be lower-cost areas that start-ups can profitably mine.

Is there a way to break free from this box? In theory, there is: by working smarter, by using available information technologies in ways that we haven't yet imagined. For Gilder, automated design tools provided a glimpse of that rational future; for others, artificial intelligence may offer the tools to build what Drucker calls the knowledge society. Such mastery of intelligent automated systems would, theoretically, lead to the self-conscious and rational control of innovation, economic growth, and productivity. It would presuppose possession of a master theory of how innovation and economic growth really operate, the fulfillment of the optimism many scientists, particularly in the social sciences, felt immediately after the war. Such mastery would, again theoretically, bring about the defeat of capital and size and produce an age in which knowledge alone mattered. Have we gotten there yet? Hardly. Certainly, mastery of information, control over knowledge, and insight are important in the evolving information industries (as they have been important for at least a century), but they have banished neither scale nor capital. To suggest that computing power has given us rational, self-conscious control is to engage in technological utopianism divorced from the grinding corporate battles of microelectronics.

That rational dream, however, has not only enlivened microelectronics. It has become, in a more immediate way, the holy grail of a field that is far less abstract and far more intimate than information processing: the world of biology, biotechnology, and drug development.

# 9

## Betting on Drug Discovery

*What one needs for success in chemotherapy is patience, financial backing and luck.*

—Paul Ehrlich

### The Trouble with Biology

In the midst of the triumphant march of physics and electronics in the 1950s and 1960s, it was easy to overlook the difficult marriage of biology and medicine. While the effects of advances in quantum physics on radar, television, and semiconductors were vividly clear, few biological advances produced much, if anything, in the way of new medicines. Although many considered the 1940s and 1950s "the golden age of drug discovery," mostly due to the development of antibiotics, vitamins, antihistamines, and the Salk polio vaccine, biology could take little of the credit. And when the pace of new drugs slowed again in the 1960s, the "failure" of basic biological research bore some of the blame.

Unlike physics and electrical engineering, biology and pharmacology, or drug development, ran on separate tracks. Research in biology occurred predominantly in academic and government laboratories such as the National Institutes of Health; research in pharmacology, which had stronger ties to chemistry than to biology, took place mostly in the laboratories of drug companies such as Merck & Co. and Eli Lilly & Co. and among academic chemists. Biologists, unlike the chemists, saw working for drug companies as a sellout to mammon; the drug industry viewed biologists as irrelevant.

Part of this friction was rooted in profound differences in methodology. As we have seen, by the 1950s, physicists and electrical engineers were working off the same quantum model, with its hard mathematical sinew and bone and high predictability. But while some biologists eagerly sought that same quantitative predictability (indeed, some physicists dropped in on biology to try to bestow it), they as yet knew so little about fundamental biological "mechanisms" that the discoveries they did make had little effect. To pharmacologists and physicians, the primary actors in the medical drama, empiricism still ruled the day. Physicians, like prewar shop-trained engineers, often prescribed regimens not because they knew why they would work but because such regimens had worked in the past or because they had shreds of evidence and shaky hypotheses that such regimens might have positive effects in the present. Medicine was, as physician and essayist Lewis Thomas put it, "the youngest science." "It gradually dawned on us," Thomas wrote about his medical school colleagues in the mid-1930s, "that we didn't know much that was really useful, that we could do nothing to change the course of the great majority of diseases we were analyzing, that medicine, for all its facade of a learned profession, was in real life a profoundly ignorant occupation."[1]

Drug therapies in the 1930s were not much better. "We were provided with a thin, pocket-sized book called 'Useful Drugs,' one hundred pages or so," wrote Thomas, "and we carried this around in our white coats when we entered the teaching wards and clinics in the third year, but I cannot recall any of our instructors ever referring to it." The first sulfa drugs in the late 1930s "were almost beyond belief," he noted, "and we sensed that the profession had changed at the moment of our entry."[2] But progress, while accelerating, was still hobbled by its empirical harness. In the drug industry, chemical compounds, which, given the advanced state of an organic chemistry enhanced by the insights of quantum mechanics, could be tailored with increasing skill, were developed and tested first in animals then in humans. If they proved effective, they were packaged and sold. Why they worked was normally beyond the industry's scope. Again, while pharmacologists could manipulate small molecules in a sophisticated way, there was no target, no larger biological context. The tools were relatively advanced, but the users of the tools were still mostly blind.

These differences produced occasional friction between biology and medicine. Vannevar Bush excluded biology from the OSRD (it seemed to him an area of pure research divorced from meaningful application) while setting up the OSRD's Committee on Medical Research.[3] The CMR successfully pioneered the development and mass production of antibiotics (notably, penicillin, which, like radar, was a wartime gift from the English), new drugs to control malaria, and new techniques to store whole blood and plasma, and it engaged in a serious investigation into the nature of shock. When Bush wrote *Science, the Endless Frontier,* he also seemed to downgrade biology, winning the enmity of biologists who actively lobbied against his NSF plan.

Biologists also had only a marginal role in the American drug industry. Until the 1930s, the pharmaceutical industry was a mass of companies, most quite small, that retained a strong flavor of its homeopathic past. As late as the 1920s, both the elite medical schools and the American Society for Pharmacology and Experimental Therapeutics barred its respective members from permanent employment in the drug industry. The depression also wreaked havoc. Between 1932 and 1934, some 3,512 small drug companies collapsed.[4] "As late as 1939, no ethical drug manufacturer in America had a sales volume as large as a department store like Macy's in New York, or Hudson's in Detroit," wrote one early postwar study of the industry. "For all 1,100 companies in the field, sales volume. . . was only $150 million."[5]

Like the division between systems houses and component makers in electronics, the prewar drug industry was roughly divided into two blocs. On one side were the ethical drug makers—Lilly, Abbott Laboratories, Upjohn, and Parke Davis & Co. were the big names—which did some research, formulated and packaged their products, and peddled them to physicians.[6] They purchased their raw materials from producers of fine chemicals such as Charles Pfizer & Co., Merck, and Lederle Laboratories (a subsidiary of American Cyanamid), plus a mix of smaller firms and mainstream chemical companies. But the development of powerful "wonder" compounds in the 1930s increased the importance of production skills, particularly in fermentation and chemical synthesis, possessed by the fine chemicals makers. Soon they found it both profitable and necessary to integrate forward into discovery, formulation, and merchandising.

As early as 1933, Merck signaled its intentions by establishing the Merck Institute for Therapeutic Research. In the mid-1930s, Merck pioneered the synthesis of various vitamins, culminating in a patented process for the full synthesis of vitamin B-12 in 1947. The company failed in its wartime bid to develop a synthetic process for making penicillin from coal tar, but it successfully concocted a partial synthesis of cortisone from cow bile and commercialized streptomycin, first discovered by Rutgers University chemist Selman Waksman (who also coined the word *antibiotic*); streptomycin was the first antibiotic effective against the great scourge of the nineteenth century, tuberculosis. For its part, Pfizer came out of the war with a major business selling penicillin and went on to develop one of the early broad-spectrum antibiotics, Ter-

ramycin. Pfizer became known for its willingness to attempt brash marketing schemes, such as sending busloads of detail men or salesmen to
hit towns across the country—which initially shocked, then converted,
its conservative brethren.[7] By the late 1940s, the traditional drug companies were fighting back by integrating backwards into processing and
production and by adopting aggressive marketing programs. An elite tier
of integrated ethical drug manufacturers soon emerged from the mass.

Although none of the top drug companies were as large as General
Electric, RCA, or Du Pont, they were growing fast, they were very profitable, and thus they had the financial power to exploit wartime opportunities. The temporary absence of very sophisticated German competition,
the boost from CMR funding, and the development of new products,
mostly antibiotics, made the American industry a worldwide force.

By the 1950s, the pace of change in the drug industry had quickened.
Between 1905 and 1935, about six new drugs appeared each year; from
1935 to 1955, thirty-seven came to the market each year; between 1948
and 1958, 4,829 new drugs, 3,686 new compounds, and 1,143 new
dosage forms were introduced. Most of these were "me-too" drugs, that
is, slight modifications of already marketed compounds. A handful were
responsible for quadrupling sales between 1941 and 1952. Unfortunately, a number of these key new products lacked strong patent protection, either because the process technology was widely shared (vitamins), because the government had sponsored the development work
(penicillin), or because licensing was open (the Rutgers Research and
Endowment Foundation freely gave out streptomycin licenses, despite
the backing of Merck). Ferocious price competition ensued.[8] In 1952,
penicillin was the largest single ethical (that is, prescription) drug,
accounting for 10 percent of the industry's $1 billion in sales, but it was a
loss leader even for the largest producers. Competition among thirteen
companies drove prices down from $2.83 per one hundred thousand
Oxford units in 1944 to $0.01 per one hundred thousand in 1952. A simi-
lar situation held for the seven makers of streptomycin, which saw prices
slide from $4.69 per gram in 1946 to $0.24 per gram in 1952. "The leading manufacturers are all operating well below capacity and foresee no
immediate solution," commented a Wall Street research report in 1953.
"Some marginal producers have been forced out of the business and
expansion by others has been delayed."[9]

These trends, so similar to semiconductor pricing, unsettled the
industry. Publicly, the big drug houses ballyhooed the same strategy of
the transistor makers: they would spend on R&D and stay ahead of
descending prices by continually developing new products. To a certain

extent that was true. By the late 1950s, the first broad-spectrum antibiotics began to appear. Unlike penicillin or streptomycin, which were "narrow range" and were thus useful against only a limited array of bacteria, the broad spectrums could be used against a wide variety of microbial-based diseases. Even better, Pfizer was able to take out a patent on its broad-spectrum Terramycin in 1949, theoretically providing a degree of pricing flexibility. A new era of patented products seemed to beckon.

But depending on patentability alone was a risky, expensive, and often unsuccessful endeavor. Novel drugs were rare, increasingly so once the first wave of antibiotics petered out and companies began to wrestle with chronic diseases, such as arthritis, cancer, and heart disease. Moreover, few products were truly new: both Lederle and Parke Davis had pioneered the broad-spectrum market before Pfizer barged in with Terramycin. In 1954 Pfizer patented an even more powerful, broad-spectrum antibiotic, tetracycline, further sharpening competition. By then the top companies had begun to cooperate, first on drug development and licensing (sharing the R&D and marketing burden), then on pricing, thus forging the notion of a pharmaceutical "club" consisting of only the major companies fronted by a powerful lobbying group, the Pharmaceutical Manufacturers Association (PMA).

In the early 1950s, Lilly, Abbott, and Upjohn agreed to a joint antibiotic development program. When Lilly developed a new broad-spectrum antibiotic, ilotycin, in 1953, Abbott and Upjohn licensed it, thus quickly expanding its market through aggressive promotion. Charges were made that only members of the club would license to each other, thus preserving their hegemony, a claim that was probably true: small companies lacked the market clout licensers sought. And without that clout, companies had to fall back on the hit-or-miss methods of drug research. Given the reality that drug discovery was in many ways a game of probability, smaller companies with smaller R&D budgets had far greater odds than larger companies of ever finding a product that could propel them into the big leagues. It seemed, in some quarters, to be unfair, raising the specter of monopoly, patent abuse, and the bureaucratic stifling of innovation. And it brought on the forces of regulation.

## Regulating Innovation

John Blair first met Estes Kefauver during World War II when Blair testified before the Tennessee congressman's Small Business Subcommittee. They discovered they were of remarkably like minds. Kefauver was, like his mentor Wright Patman, a vigorous antimonopolist and fervent supporter of small business. An investigation by Kefauver into the monopoly question in 1947 produced "United States versus Economic Concentration and Power," another extension of the basic arguments of Berle and Means, and the Kefauver-Cellar Act of 1950, which discouraged horizontal mergers and inadvertently helped spur the conglomerate wave.

By the time Kefauver entered the Senate in 1953, Blair had become his economics adviser and chief investigator.[10] Described by one of Kefauver's biographers as "a morose, singleminded man who did not endear himself to his associates," Blair spent nearly his entire career either in the New Deal bureaucracy or on Capitol Hill. Although Blair recognized the centrifugal forces of decentralization and renewed competition, he also believed that the government had to help the process along. To Blair, as to Kefauver, the issue of monopoly turned on the question of pricing. Drawing on the work of Gardiner Means, Blair and Kefauver focused on what they viewed as a plague of administered pricing—prices set by companies independent of market forces—which they identified, again following Means, as a key factor behind inflation. In successive investigations from 1957 to 1962 (bracketing a losing bid for the Democratic presidential nomination in 1960), the Kefauver-led Antitrust and Monopoly Subcommittee launched probes into the automobile, steel, baking, and drug industries and, in a series of dramatic hearings, branded each guilty of administered pricing.

The Kefauver hearings climaxed when Blair and Kefauver took aim at the pharmaceutical elite.[11] Blair's staff produced evidence that the major drug companies were using monopoly power from patents to set prices well above costs, that different companies charged the same price for similar drugs, that prices overseas for U.S.-made drugs were lower than those made at home, and that firms were buying drugs in bulk from suppliers and selling them at higher prices. The hearings also raised the question of safety and warned of the dangers of false advertising (the first hint of birth defects caused by the tranquilizer thalidomide surfaced at a Kefauver hearing). Kefauver concluded that the major drug companies did not effectively pass the benefits of research along to consumers, in the form of either lower prices or innovative therapeutics, and that monopoly power and size in pharmaceuticals not only did not nurture innovation but actually retarded it. "From this, it is obvious that the public does not benefit by the applied research of the pharmaceutical industry," said Kefauver. "It is also obvious that the selling price for a particular ethical specialty is not predicated on the cost of materials, but, rather, predicated on what the traffic will bear."[12]

In 1962 Kefauver drafted a drug control bill that sought to slash patentability from seventeen years to three years, set up a federal licensing system for all drug manufacturers, require that generic names be displayed on labels and in advertising, and force drug companies to demonstrate both safety and efficacy before putting new drugs onto the market. Kefauver's bill—particularly its patent provisions—soon bogged down. In the end, only the safety and effectiveness provision survived, and only after Blair helped publicize the damage wrought by thalidomide in Europe by leaking the story to the *Washington Post*.

Although the 1962 Kefauver-Harris amendment to the Food, Drug and Cosmetics Act was the most important pharmaceutical legislation of the postwar era, the drug industry still could feel that it had escaped real punishment. By retaining full patent protection, the industry preserved its ability to set prices freely. Despite claims that increased testing for Food and Drug Administration approval would lead to research cuts, over the long term basic research was probably not harmed, particularly with the federal government supporting an expansionary biomedical research establishment. Besides, profits from new, major products were high and pricing on many patented drugs was inelastic, thus allowing the companies to set prices high enough to stay ahead of rising regulatory costs. Ironically, the safety and effectiveness provision of the bill only raised the barrier of entry to new competitors and, over the long run, strengthened the position of the largest companies.

The original Kefauver bill was an aggressive attempt to reform the structure of a research-based industry, to channel innovation and competition. The final bill ignored Blair and Kefauver's primary concern, but the pair still succeeded in planting serious doubts about the nature of the pharmaceutical enterprise. Both Kefauver and Blair believed that the technological engine powering the drug industry was operating at less than full efficiency and that it had become that way because of the extraordinary freedoms afforded by patentability. At the heart of Kefauver's bill were strongly held assumptions about the nature of drug discovery and the relationships among corporate size, innovation, and capital. The bill raised fundamental questions, which continue to hang over any attempt to create optimal conditions for corporate innovation. How much effect does the underlying science have on the structure of the industry? Is there an optimal patent policy for a particular stage in the life cycle of a technology? What is more important to consumer welfare: new, improved products or lower prices? How much profitability is enough to ensure long-term investment in R&D? How far should the government regulate?

The Kefauver attack forced the industry, and its allies, to develop a rationale for the way it did business; the persecution of the drug industry became a favorite subject of the free-market economists of the Chicago School. When they looked at pharmaceuticals, they saw a highly fragmented industry in which no single company held overwhelming market share, though they acknowledged that companies did dominate specific therapeutic classes. They argued that competition did exist, though less on pricing than on the development of new drugs: competition through innovation. Given the high costs of drug development and the uncertain nature of biological knowledge, flexibility for patents and pricing was the reward for shouldering the risks of research. Any reduction in patentability, any attempt to narrow profits by holding down prices, would result in smaller research programs, fewer new drugs, and diminished consumer welfare. Finally, drug development at large companies did produce new, effective "wonder drugs." The underlying economics of drug development demanded a certain size and financial power.

The two views of drug research and drug economics were thus radically at odds. Kefauver and Blair saw a bureaucratic industry gouging the public, exaggerating its costs, exploiting the patent system, and developing new drugs at a deliberately leisurely pace. The free market analysts saw a fit, modern, aggressive industry engaged in tortuous basic research and using price inelasticity to support very long-term, very risky projects. Both sides thought they were preserving the free enterprise system— Kefauver and Blair by reviving competition, the free marketers by allowing the competition that existed to work its magic. Where was the reality? In fact, it was lost somewhere in the middle. The dynamics of the drug industry demonstrate particularly clearly the subtle, if powerful, relation between the state of an underlying science and the structures visible in the marketplace.

## The Gentlemen's Conspiracy

*Although competition became cutthroat, the industry's leaders never forgot they were gentlemen when it came to prices. The price war that turned penicillin and streptomycin into distressed merchandise in the late Forties was an experience that left its mark. Since then the drug companies have been extremely loath to use price cutting as a competitive weapon.*

*—Fortune, August 1965*

In retrospect, the drug industry's pattern of evolution in the 1940s and 1950s reprises that of the rise of the giant industrial corporations of the late nineteenth century. First there was a fairly primitive, decentralized, atomistic industry; then came expansion, rapid consolidation, and shake-out driven by technological change; ruthless competition occurred next, followed by an attempt (so similar to the efforts of J. P. Morgan, Sr.) to bring order out of chaos. Finally, of course, came the expected counter-action by the government in the form of increased regulation.

Such a pattern would be less convincing if it depended strictly on the evidence of administered pricing, favoritism in licensing, and a failure to innovate gathered by Kefauver and Blair. In fact, evidence of more bla-tant attempts to control prices and markets also surfaced, and it could not be so easily dismissed as a matter of philosophical differences. Kefauver and Blair had sketched out a broad pattern of "sticky" prices within the club, but it took the Federal Trade Commission to produce a smoking gun. In 1958, 1963, and, after an appeal, 1966, the FTC brought charges against what it described as a cartel made up of Pfizer, Squibb, Upjohn, Lederle, and Bristol-Myers that had formed in the early 1950s to fix prices on broad-spectrum antibiotics.[13] Pfizer, Lederle, and Bristol-Myers, the FTC argued, had been involved in the scramble to patent tetracycline in 1954, but rather than slug it out, they chose to carve up the market instead. By secret agreement, Pfizer won the patent, while Lederle and Bristol-Myers (with its licensees, Upjohn and Squibb) received licenses. The five companies then agreed to fix prices. In a 1965 Senate speech, Louisiana senator Russell Long brandished corporate memos describing "pow-wows" (price-fixing meetings) and "sinners" (those who threatened to upset the pricing order) and letters discussing concern about upsetting the pricing order.[14]

Particularly damning was the pricing data.[15] From 1951 to 1964, prices on various broad-spectrum antibiotics were nearly identical. The FTC began a probe in the late 1950s, and in 1963 it ruled that patent deception and price fixing had taken place and ordered Pfizer to license the drug to all comers for a 2.5 percent royalty (more, by the way, than Bristol-Myers and Lederle were paying). And indeed, by 1964, prices had begun to fluctuate. Squibb, with the smallest market share of the group, dropped its price dramatically from $22 to about $4, despite earlier internal memos that showed it fearful of getting a reputation in the industry as a "cost cutter." Lederle, the market leader, only slowly and gradually lowered its price, again after the FTC probe. Even while the FTC was suing the group, Pfizer was suing McKesson & Robbins, a drug wholesaler, for bringing in tetracycline from overseas and selling it at one-third the cartel price.

The FTC charges provide clues to how the industry reacted against the unrestrained competition of the late 1940s. Technological progress had begun to accelerate (though major innovations still appeared slowly), far more than in other advancing industries, and the drug industry could not really explain why. Industry leaders lacked the tools to give them the certainty that innovation would continue. It was a matter of grave anxiety—and the industry, dominated at the top by marketing men, moved to alleviate it.

The notion of the drug industry as a sort of throwback to the late nineteenth century operates on several levels. Drug discovery was an industrial process as much as steelmaking or automobile manufacturing; in fact, nearly all of Kefauver and Blair's targets were industries driven by mature industrial technologies. The drug industry had few belching smokestacks and hordes of Druckerian white-collar knowledge workers, but its major output—new pharmaceutical products—was still determined by the same economies of scale of any steel mill or car plant. The reason so many observers, including Blair, missed that point is that these were economies of scale not of production but of R&D. Drug companies did not primarily study disease, they formulated thousands of chemical compounds annually. Pharmacologists did not labor to understand how a compound and a disease interacted—how could they, given the sketchiness of biological knowledge?—they sought to test as many compounds as they could to see what worked; the more money spent, the more compounds screened, the greater the probability of discovering the next penicillin. Advances in chemistry helped. The drug industry was producing more new drugs not because it better understood underlying biological reality but because it had learned to manipulate small molecules chemically and to organize screening more efficiently. Unlike electronics, there was no sense, hubristic or otherwise, that drug discovery was self-sustaining; and, not surprisingly, there were as yet no pharmaceutical Sarnoffs and Marshalls to sketch a brilliant future.

The rise of big drugs as a sort of latecomer to a mature industrial economy explains the industry's insecure mix of modernity and archaism, collusive activities and free market posturing. The companies were better at performing a range of well-defined, specialized functions (a kind of research Taylorism), than at embarking on undefined, uncharted research programs; no drug company felt as confident of the possibilities of medical research in 1960 as Du Pont had in polymer research when it built Purity Hall in 1926. Creativity and innovation, key elements of basic research, simply did not pay; given the incomplete state of biological knowledge, mass screening was an alternative that was both cheaper and safer. Moreover, once the initial discovery was made, the drug companies could use their chemical skills to modify, improve, and, of course, extensively test. Organizational structures were thus designed for redundancy and process more than innovation.

Pharmaceutical R&D was a classic bureaucratic process. The case could be made that the bureaucratic mode was not only a natural response to organizing an industrial process like drug discovery but an effective one. Again, bureaucracy—management through set routine— was forced upon the industry by the empirical nature of drug discovery and the sketchiness of the underlying biology. Mass screening, despite the creative work that certainly accompanied it, was essentially a mechanical process of set routines. Like the post office moving letters or an insurance company processing claims, the drug companies shuffled through hundreds of thousands of compounds every year. Even if the FDA had not existed, paper still would have proliferated. The process of screening and testing produced a vast river of data that had to be recorded, filed, collated, weighed, passed along, acted upon. Such functions require a bureaucracy.

Despite their adversarial relationship, the major drug companies and the FDA mirrored each other: highly centralized, hierarchical, full of pharmacologists, biochemists, chemists, physicians. In both cases, the mandate, particularly after 1962, was to limit risk. The FDA was slow in part because of the "bureaucratic" tendencies of any large regulatory body, particularly one whose mandate is to protect consumers from harm, not to ensure industry profits. The companies, which were essentially risk averse, were slow as well. Certainly they would take on some high-risk ventures, but their potential for failure was always calculated. Insulated by their long-term patents, confident that the pace of change was not accelerating and that the probabilistic calculus of mass screening would predominate, they could erect organizational structures that double checked and triple checked, that reviewed and recalculated at every stage of a long commercialization process. Both manufacturers and regulators were thus designed not for great leaps of thought but for careful progress. Both manufacturers and regulators were set up to keep information flowing through a number of specialists. And both were defined not only by internal dynamics but by the contingencies of the underlying science.

The failure to penetrate the realities of drug discovery created some of the misguided aspects of the cases made by Kefauver and industry defenders. Kefauver and Blair's plan to drop patentability to three years probably would have forced most of the companies out of ethical research; although the industry probably exaggerated R&D costs, a three-year exclusive would almost certainly not have been enough to earn a reasonable return on capital even if, as Blair wanted, much of the marketing budget was eliminated. Meanwhile, increased regulation would further exclude smaller companies by raising barriers to entry. Blair also thought it highly revealing and damning that many of the greatest pharmaceutical breakthroughs came from individual inventors, mostly university chemists, not from "bureaucratic" corporate research teams (he ignored the fact that corporate money supported many of these investigators, as Merck backed Waksman as well as the development of the corticosteroids).[16] Blair's view of innovation and invention could also be very simplistic. From the drug industry's point of view—and it was a perfectly commonsensical one—invention *was* an exaggerated skill. Drug discovery was *not* rational; it resembled, as one observer said, "molecular roulette."[17] Given that reality, companies leaned on the two assets they could control: financial power and a complex organization—two assets every academic department, lone inventor, and small company lacked. In 1976 economist David Schwartzman accurately characterized the continuing challenge of drug development:

> Drug discovery today requires an inordinately expensive and highly organized multidisciplinary research team effort, well beyond the scope of academic laboratories which have neither the resources nor the facilities or temperament to indulge in the development of new drugs. A case in point is the development of the antibiotic field. An enormous effort was made... over a span of twenty years in a search for new antibiotics. Yet the practical outcome of this mammoth effort was the discovery of a mere handful of antibiotics which found clinical application. From the financial point of view alone, such a costly undertaking, resulting in relatively few winners, could only have been possible in the goal-oriented atmosphere of the industrial laboratories.[18]

Under such conditions, the industry not unexpectedly strived for control. Discovery was not predictable, and companies often felt they had little say over where or when it would occur. But they did control great financial resources, a function of monopoly profits from patented drugs, which they could use to dominate the commercially relevant work that serendipitously arose outside their own laboratories. This helps explain the industry's steady carping about federal spending on R&D, particularly biology in the 1950s, which in theory posed a competitive threat to their own hegemony.

But while Blair's view of drug research lacks context, Schwartzman's picture amounts to an apologia. Schwartzman, who received financial help from Pfizer and technical assistance from a number of drug companies, rationalizes the pricing patterns but sidesteps the collusion issue. He labors hard to prove the Schumpeterian thesis on size and innovation, a rationale for retaining the recently established order.[19] Drug companies are not bureaucratized, he maintains, because there are few layers between top policymaking management and research (he ignores the research bureaucracy). Firms take great risks, spending money on projects they have no certainty will ever pan out. Just because it is difficult to isolate individual discoverers of drugs does not mean that the research was not basic. Like Frank Jewett and William Shockley, Schwartzman dismissed the distinction between discovery research and developmental work by saying that such terms are not very useful in empirical pharmaceutical research (he was right). And if innovation slowed in the later 1960s, he blamed the FDA for forcing a clinical testing regimen on researchers that "bogs down the entire progress of drug discovery because it leaves the scientist without a facile indicator of how his work is measuring up and therefore leaves him unable to identify those critical leads that are demonstrably relevant to clinical therapy." In other words, if the FDA were not always asking questions, the companies could screen more compounds and come up with more new drugs. That is true, of course, but at the danger of poisoning the market with either useless or occasionally dangerous products. Markets would learn, of course, but at the cost of human sacrifice.[20]

Schwartzman's central argument, and that of the PMA, was that drug regulation was making innovation too expensive to pursue. He claimed that the diminishing number of new drugs approved was a direct result of the 1962 amendments; he rejected the arguments that as medical targets became more elusive, the research task became far more difficult, and that increasing costs had forced companies to clean up their portfolios and focus only on compounds that appeared clearly useful. Schwartzman's calculations that, by the mid-1970s, the return on R&D had plummeted to 3.3 percent after taxes, down from 12 percent in the 1960s, received widespread publicity as the debate, following the political winds of the late 1970s, shifted from pharmaceutical collusion to FDA obstructionism. "While not all of the sharp drop in drug innovation can be laid at the doorstep of regulation, economists argue that by ignoring the economic impacts of regulation the FDA is having a devastating effect on the drug industry," declared Business Week in 1977.

Devastating? Not if you look at the industry's results from those years. With pricing freedom intact, and despite a downturn in new-product introductions, the drug industry cruised through the turbulent 1970s and into the 1980s. Even pro-industry voices, such as Sam Peltzman at the University of Chicago, who first demonstrated the increased cost of regulation on R&D expense, and FTC economist Frederic Scherer, cast doubts on Schwartzman's figures and his tendency to extrapolate from past levels of profitability. "Either the drug companies are stupid or they know something Schwartzman doesn't," said Scherer.[21] In fact, the combination of increased regulation and pricing freedom gave the big companies a weapon against smaller competitors—as long as the prevailing empirical R&D orthodoxy held. By lowering regulatory barriers, the club would have invited many small competitors to play. Heavy regulation, in other words, was a conservative force. Studies show that between 1957 and 1961 the four largest companies controlled about 24 percent of the "innovational output"; between 1967 and 1971 that had risen to 48.7 percent.

By the late 1970s, the industry radiated ample, second-generation wealth.[22] Profits were high and seemingly recession proof; the companies built architecturally striking facilities, many of them across a belt in northern New Jersey. While Merck and Lilly were often mentioned as growth stocks, and awarded growth multiples, the members of the club lived in happy isolation from the riot of the market. Cash flows were so strong that research spending, about 9 percent of sales, and marketing were easily supported out of retained earnings. Share prices were high enough that with the exception of Squibb, acquired by chemical company Mathieson in 1952 (merging into Olin Corp. two years later to form Olin Mathieson), the conglomerates never posed a threat; Wall Street could hardly complain about pharmaceutical performance. The small amount of merger activity that did take place occurred within the industry. Perhaps the most striking aspect was the industry's stability, so unusual in technologically sophisticated postwar industries, though significantly very similar to many prewar industries. "The sources of innovation are declining," declared one economist in 1977. "With the cost of developing a new drug soaring, research is a game smaller companies cannot afford to play."[23] If that was the case, it certainly did not noticeably ruffle the largest companies.

Despite the prosperity, the industry confronted a threatening world. Along with many other industries, pharmaceuticals actively diversified in the mid-1960s, reinvesting their excess cash not only in more ethical drug research but in new business lines such as cosmetics, animal health, chemicals, even confections—none of which were as profitable as drugs (though they were certainly more stable and predictable). The companies rationalized this shift by blaming costs that resulted from the Kefauver amendments, as well as fear of new regulation. More threatening, though less visible, was the possibility that either the R&D orthodoxy would give way or Congress would act to shorten patent lives or control drug prices. And with Wall Street increasingly fascinated by smaller technologically based companies, even the industry's great weapon, its financial power, no longer appeared to be as dominant as in the past.

## Syntex: The Rational Dream

The notion of an ancient pharmaceutical elite was a myth, more fantasy than fact. The power of the drug elite was self-made, relatively recent, and dependent on a small basket of products. "The PMA club" did not coalesce in its modern form until the 1950s, around the time Syntex, the only new member to invade its ranks since World War II, was founded in Mexico City.

Syntex was, for a time, very different from the rest of the PMA companies. In its most creative days, the company embodied many of the centrifugal forces Blair saw in other industries but missed in pharmaceuticals. Syntex exploited a series of newly discovered, natural compounds, the steroids, which promised unlimited therapeutic potential and seemed to suggest that a homogeneity existed beneath the welter of human ills. Financially, Syntex not only found itself part of a conglomerate for a time but tapped the bubbling optimism of the 1960s stock market to fund itself, leveling the financial playing field against powerful competitors. Most significant, Syntex depended upon its mastery of a sophisticated chemistry, backed by sketchy, if still substantial, biological research. The company went at least a half-step beyond mass screening toward truly rational, or predictable, drug discovery, toward drug engineering as opposed to drug discovery. Far more than any example of Blair's, Syntex represented the innovative potential of a small, entrepreneurial, scientific operation over a large, empirically rooted establishment. Syntex was proof that R&D economies of scale, buttressed by pricing freedom, might not rule forever.

The Syntex story begins with a mysterious steroidal hormone called cortisone, a regulatory chemical produced by the adrenal cortex, two small glands perched atop the kidneys. Researchers established the first link between the adrenal cortex and hormonal activity in 1930; five years later, they isolated the first small quantities of adrenal steroids. But it was only during the war, when rumors (unfounded) surfaced that German pilots had been given cortisone extracts allowing them to fly without oxygen at 40,000 feet, that the National Research Council organized a U.S.–Canadian effort to synthesize cortisone and other steroids.

Merck's Lewis Sarett developed a thirty-six-step partial synthesis for cortisone out of cattle bile, the most complex production process of its day. Merck's cortisone was used in tests by Dr. Philip Hench of the Mayo Clinic that showed such remarkable results against rheumatoid arthritis that cortisone was quickly dubbed a "miracle" drug.[24]

Soon, cortisone was tested with dramatic results against gout, rheumatic fever, lupus, allergies, ulcers, lymphatic cancers, even common, everyday fatigue. But while the drug showed a remarkable ability to relieve symptoms of diseases, it did not cure them; as soon as the treatments stopped, the maladies returned. Nonetheless, in language that would echo with other future "breakthrough" therapeutics, *Fortune* in 1951 wrote about cortisone as if it were some all-controlling metachemical: "As therapeutic agents, the hormones differ from nearly all previous drugs. They are not germ killers. . . but directly alter the complex glandular balances of the human body. But few doubt that, in the continuing story of hormones, medicine will soon be striking at some of the most deep-seated organic complaints of man."[25] Even sober *Scientific American* succumbed. The magazine carefully pointed out how "universal remedies" have persisted even though "scientific medicine has consistently frowned on all unitary theories," yet it then dismissed such doubts in this new steroidal age:

> Despite the fragmentary nature of our knowledge of how cortisone works, it is clear that the pages of history have turned a new chapter in man's long search for the mechanism of disease. Thus far we have only been able to read a few disconnected sentences of that new chapter. But they are so amazing in what they tell us and so revolutionary in what they imply that the medical world today is waiting anxiously, one might say breathlessly, for the next development.[26]

The next developments, however, depended on the production of large quantities of raw steroids. Enter a lone, eccentric professor named Russell Marker.[27] In the late 1930s, Marker taught chemistry at Pennsylvania State College and researched natural precursors of steroidal compounds. Like many drug hunters, Marker's modus operandi was to tramp through forests and fields looking for natural steroidal sources. (This was common corporate practice too: Bristol-Myers once sent shareholders little envelopes with their annual reports and asked them to send back dirt samples. When the cortisone race heated up, five individually sponsored expeditions were sent to Africa to seek out one rare vine.) In 1939 Marker, supported by Parke Davis, discovered that he could extract quantities of a steroidal precursor called diosgenin from the root of the sarsaparilla plant; he was then able to turn diosgenin into progesterone, a sex hormone that prepared the uterus for the implantation of the egg, helped to maintain pregnancy, and prevented multiple fertilizations from occurring. Marker published a paper in the *Journal of the American Chemical Society* and then continued his search. For even the diosgenin produced by the sarsaparilla proved inadequate for the needs of researchers and commercial manufacturers. Then Marker heard about the lumpy, black root of an inedible hillside yam that grew in Mexico: *la cabeza de negro*.

In 1943 Marker rented a small laboratory in Mexico City. A year later, he appeared at the offices of Laboratorios Hormona, a local marketer of raw, natural steroids, hugging two jars wrapped in Mexican newspapers containing 4.5 pounds of progesterone worth about $160,000. That impressed one of the company's co-owners, Dr. Emerich Somlo, a Hungarian refugee who had settled in Mexico. Somlo had been buying small amounts of very expensive naturally derived hormones and then reselling them, mostly to a worldwide hormone cartel dominated by Merck and Ciba, the Swiss drug giant. Now Marker, off the street, was hauling around enough to equal the world production. Somlo immediately offered to fund Marker's work through a new company called Syntex S.A. and share the profits. Marker agreed and went back to his yams and the five-step partial synthesis process he had developed. In 1944, with war raging in Europe and the Pacific, he produced several more pounds of progesterone. Then, in early 1945, he argued with Somlo and stomped off, confident of his own indispensability.

Marker miscalculated. In Havana, Somlo found George Rosenkranz, a twenty-nine-year-old fellow Hungarian who had studied with Swiss Nobel Prize–winning chemist and steroid pioneer Leopold Ruzicka. Rosenkranz had read Marker's earlier papers and thought he could retrace his steps and produce diosgenin from the *cabeza de negro*. Somlo convinced him to give it a try. Within a few months, he had succeeded in emulating Marker's technique; within a year he had developed a full synthesis of testosterone, the male sex hormone. Within five years, Syntex developed the means to produce synthetically all four major groups of steroids—androgens, estrogens, corticoids, and progestogens—from diosgenin. Syntex's development of the first synthetically produced cortisone in 1951 marked the climax of the cortisone boom and put the company on the research map as well as in the pages of *Life* (where Syntex researchers are pictured standing behind a huge yam), *Newsweek, Harper's,* and *Business Week. Harper's* particularly noted the issue of size. "As perhaps to other recent development, it [cortisone] also underscores a point often overlooked in a big-money age. Big minds, rather than big research budgets lead to big discoveries."[28] As for Marker, he tried to set up a rival diosgenin company and failed.

By then, Syntex had broken the hormone cartel and was selling bulk steroids to American and European drug companies. Rosenkranz, however, had bigger plans. In 1950 Gregory Pincus of the Worcester Foundation for Experimental Biology showed that he could prevent rabbits from ovulating if he injected them with progesterone. Dr. John Rock, a Harvard gynecologist, then confirmed the finding in humans. Pincus was a consultant with G.D. Searle & Co., an Illinois drug company, and he convinced Searle to try to develop a more powerful synthetic steroid. At the same time, Carl Djerassi, the wunderkind who had led the Syntex team that synthesized cortisone, now turned to the problem of synthesizing a potent progesterone from a diosgenin base. Djerassi's team came up with a hormone called norethindrone, which proved to be a far more potent version of ordinary progesterone and, when taken orally, seemed to inhibit ovulation. Not long afterward, Searle announced a similar material, norethynodrel, with much the same potency and effect. Both were tested at Pincus's clinic, and suddenly the idea loomed that a birth control pill might be feasible.

How did a group of such talented chemists end up with an obscure Mexico City drug supplier? Syntex had profited from the disruption caused by World War II. Central Europe had been a thriving center of chemical research before the war, and many of its finest minds were subsequently scattered. Syntex offered a neutral site—Mexico was not involved in the war—and a decent living for someone like Rosenkranz. Djerassi, on the other hand, had arrived as a teenager in America with his Viennese mother, a physician, in 1939. After stints at Kenyon College, the University of Wisconsin, and the U.S. unit of Ciba, Djerassi took up an offer from Rosenkranz to come to Syntex. In his memoirs, Djerassi recounts that decision. He had never heard of Syntex and always thought that "serious chemistry stopped at the banks of the Rio Grande."[29] Rosenkranz, then thirty-two, charmed him both personally and professionally; the offer involved a full laboratory, an associate directorship, and interesting work. Djerassi's real dream was to have an aca- demic career, which he thought he could launch if he could author enough scientific papers. "I felt intuitively that this was the right place. Syntex had the same objective I did: to establish a scientific reputation. Our common goal—a new and more productive synthesis of cortisone from raw plant material—was one of the hottest scientific topics in organic chemistry at the time."

Rosenkranz and Djerassi, with their strong academic backgrounds, established a different approach at Syntex from usual empirical drug practices. Djerassi considered himself a medicinal chemist and recognized how much luck and mechanical screening contributed to most pharmacologic advances.

> It is hardly surprising that the modern medicinal chemist is unhappy with this state of affairs, as predictability rather than serendipity is the essence of science, and especially of chemistry. Chemists such as Paul Ehrlich, who founded modern chemotherapy in the early part of the century, have attempted to establish relationships between chemical structure and biological activity that lead to the a priori prediction of a potentially useful drug. To a considerable extent, the development of steroid oral contraceptives represents a successful instance of this predictive approach, in which we deliberately set out to synthesize a substance that might mimic the biological action of the female sex hormone.[30]

In short, since progesterone's chemical structure was known, chemists only had to devise a means to alter a naturally occurring variant so as to "mimic its behavior." The trouble was, of course, that there were few such substances known—particularly when it came to the vast number of naturally occurring macromolecules in the body. Although the predictive or rational approach worked in synthesizing a number of steroids, it was not as successful when it came to *using* many of them; the birth control pill was a great exception. Indeed, while the action of progesterone was fairly well known through empirical studies, the intricate interactions of other steroids with various biological and chemical systems were not. Neither cortisone nor any of the other hormones proved to be the master regulator, and their serious side effects limited their use over long periods of time—a story that would be repeated again and again over the following decades.

Still, Syntex's fleeting grasp of rational drug discovery was enough to offer many commercial advantages. Translating these advantages, however, was not easy. Without the resources to set up an American marketing operation, Syntex licensed norethindrone to Parke Davis, as a payback for supporting Marker's work, and began selling some of its steroids in bulk, particularly for skin ailments, in the huge American market. In the long run, both initiatives put Syntex at the mercy of larger competitors. Parke Davis moved slowly to get norethindrone approved by the FDA as a palliative against menstrual difficulties; returns were small. As a bulk supplier, Syntex sold a steroid to Merck, which would sell it by prescription for as much as ten or twenty times cost. High prices kept markets small. As time passed, other companies, including Merck itself, came up with efficient synthetic processes for making bulk steroids, forcing prices down. Syntex was caught in the same trap as other small drug companies. After making a profit of almost $2 million in 1953, Syntex barely broke even in 1956. What the company desperately needed was to follow the path taken earlier by Merck and Pfizer: integrating forward while producing a stream of new products. However, Merck in the 1930s had the financial resources to fund such an expansion; Syntex did not.

Syntex needed a partner. In the mid-1950s, some representatives of the investment banking firm of Lehman Brothers in New York poked around, but they were distressed by the company's location and its confusing books. In Lehman's wake came a far less orthodox financier by the name of Charles Allen. A grade school dropout, "Charlie" Allen had gone to Wall Street as a runner in the 1920s, then took up investing. The crash wiped him out, but he then set up a private investment house called Allen & Co. with his brother, Herbert, and started again. By the 1950s, Allen had become something of a legend.[31] He operated with stealth and secrecy outside the traditional web of investment banking relationships. He invested for his own account, hiring younger men purely on their commercial, money-making abilities. Over the years, he had made large, very lucrative investments in Hollywood, in the Bahamas, in a company called Benguet built around a Philippine gold mine, and in a dozen other special situations. Allen had read about Syntex in *Fortune* and was interested, though he knew little about science.

In 1956 Allen bought Syntex and folded it into Ogden Corp., a nascent conglomerate he controlled. In June 1957 Syntex Corp. was organized as a Panamanian company that within a year or so bought out the original Syntex S.A. By 1959 the company was losing half a million dollars a year because of the costs of its research program, which was just reaching its most productive, and expensive, period, and of setting up the U.S. subsidiary. To support those expenditures, Syntex signed a five-year research deal with Eli Lilly in 1959, in which the Indianapolis company, the second largest U.S. drug company behind Merck, would pay half the cost of Syntex R&D in exchange for comarketing rights on any new product. The Lilly deal, like its scientific reputation, built credibility for Syntex on Wall Street.[32] In April 1958, Allen announced that he was spinning part of Syntex off to the public; he offered some 1.2 million shares of the company, giving Ogden shareholders the right to buy one Syntex share for every four of Ogden they already owned. After one hundred thousand shares went unsold, officers and directors were able to pick them up at a mere $2 a share.

Allen's timing was impeccable. Syntex opened at $27.50 on October 13, 1960, on the American Stock Exchange, selling 80,000 shares. The birth control pill was finally taking off—though it had not been easy. In 1957 Parke Davis and Searle both received approval for their versions of norethindrone to relieve menstrual problems. At that point, Searle began the testing needed to get FDA approval for using the steroid as an oral contraceptive. Parke Davis, however, retreated, fearful of a Catholic boycott, forcing Syntex and its licensees to repeat some of the clinical testing (Pfizer had taken an option on norethindrone from Syntex a few years earlier but also backed off because of the scruples of its Catholic chairman, John McKeen). Syntex instead signed up Johnson & Johnson's Ortho division, which won approval to market the drug as Ortho-Novum in 1962, two years after Searle had done so; it quickly became the largest selling birth control pill in the world. By then, Parke Davis had changed its mind and gained FDA permission to market its own pill, Enovid, in 1964, the same year Syntex began to market its own, Norinyl.[33] By the mid-1960s, Syntex was responsible for producing four of the five U.S. birth control pills.

Syntex's step-by-step climb is a classic model of forward integration. Rosenkranz expanded the company in discrete stages, each perilous. First it sold bulk steroids to customers outside Mexico, mainly in America. Pouring its money back into research, Syntex developed new products, including the birth control pill. With cash from steroid sales, the Lilly deal, and the sale of stock, it was able to advance in two directions: setting up a marketing group in America and beginning to manufacture, package, and seek regulatory approval for products under its own name. The push was always to eliminate the middleman and capture as much of the profits as possible. Finally, knowing that its birth control pill patents would eventually lapse, Syntex, then based in Palo Alto (though retaining a Panamanian registry for tax purposes), began to pour its cash flow into new steroidal therapeutics and new business lines.

For all its skills, Syntex would never have broken into the club without the stock market. Syntex was a favorite growth stock of the 1960s. From a low of 11 in 1962, the stock peaked at 260 in 1963, triggering a three-for-one split, before sliding to 213 on November 22, 1963—when trading was halted by the Kennedy assassination. When trading resumed the following Tuesday, the stock started out at 63 and roared to 135 (495 without the split) before falling to 125. Rumors soon spread: Syntex is cutting prices, steroids cause cancer, insiders are dumping their stock. The stock sank. Then, the equal and opposite reaction: Syntex is announcing new products, steroids cure cancer. Up the price flew.

No one pretended that the perspective of the public markets was sophisticated scientifically. The stock market overestimated the "revolutionary" impact of Syntex R&D and thus its ability to develop new products in the future. Success in synthesizing cortisone, progesterone, and other hormones did not necessarily translate to success in developing other compounds. In the decades ahead, while Syntex regularly spent more as a percentage of sales than any other drug company—though not, of course, in gross terms—its R&D productivity slowed. Although the birth control pill was a success (despite intermittent charges that it caused a variety of ailments), many other uses for steroids fizzled. Through the 1980s, Syntex made the bulk of its drug profits from Naprosyn, a so-called nonsteroidal antiarthritic. Although Syntex successfully joined the pharmaceutical elite, by the late 1980s it was still the smallest major U.S. drug company and thus always at a disadvantage, particularly when it came to spending on R&D programs that for the most part remained stubbornly based on empirical processes and driven by scale economics.

## The Threatening Environment

By the 1980s, the environment around the drug business was growing more difficult. Ballooning overheads drove firms to sell globally (to extract the greatest return from a given product), which drove costs up even higher. A cost spiral soon developed, triggered by a combination of rising R&D expenditures and lagging research productivity—that is, a dearth of major new products. The presence of so many drugs going off patent, plus a heightened awareness of rising health care costs, fed a growing generic drug industry. To generate new, innovative products, the major companies found that they had to fund both mass screening and new rational discovery programs (to do so, many companies sold unrelated assets, reversing two decades of diversification) and engage in increasingly expensive clinical testing programs. The benchmarks were set by Glaxo Holdings and Merck, which spent respectively 12 percent and 14 percent of sales on R&D—around a billion dollars each in 1992. The need to recoup those costs turned pharmaceutical competition from plodding, gentlemanly affairs into swiftly spreading world wars. The classic new-product campaign was waged by London-based Glaxo, whose product, the anti-ulcer agent Zantac, was aimed at toppling what had reigned for some time as the largest selling drug worldwide, SmithKline Beckman's Tagamet. Glaxo opened the campaign in the early 1980s, when it blitzed Tagamet, first in Italy (a traditionally strong Glaxo market), then in one international market after another, culminating in the United States.[34] Glaxo's Zantac was not demonstrably superior to Tagamet, but Glaxo's marketing certainly was, with its massed and orchestrated strike forces of salespeople (often supplemented by "renting" sales forces from companies that had few products). Tagamet and SmithKline were in trouble even before Tagamet lost its patentability; although the drug still generated $1 billion in sales by 1990, Zantac pulled in over twice that. In 1989 SmithKline merged with English drugmaker Beecham.

These were economies of scale with a vengeance. And yet, ironically, even as members of the drug elite were entering a period of consolidation—the most dramatic since just after World War II—the financial markets were investing in masses of new competitors that called themselves biotechnology companies. These new companies noisily claimed to possess the secret of rational drug discovery and argued that they could vault the barriers of R&D scale and marketing reach. There had not been such a sharp collision of two opposed groups of companies since the transistor makers confronted the vacuum tube oligopoly. But if biology is different from physics, so too was the first decade of biotechnology different from the transistor wars of the 1950s. And those differences again stem from the perils and pitfalls of rational drug discovery.

# 10

## Biotechnology's Incomplete Revolution

### The Grail of Molecular Biology

Although dominated by chemists, the small group responsible for building Syntex grasped the central importance of biology. Djerassi left Syntex in the early 1950s for Wayne State University, then returned to Mexico City to head up Syntex's R&D program in the middle years of the decade.[1] In 1960 he was recruited by Terman, by now Stanford's provost, who was eager to build a world-class chemistry department and to promote Stanford Industrial Park, with its already burgeoning network of academic–commercial ties, mostly in electronics. Djerassi, a man of unusually wide interests, was perfect for Terman on both counts. In Palo Alto, he quickly discovered the presence of a new interdisciplinary blend of genetics and biochemistry known as molecular biology. Soon afterward, he convinced Syntex to build the Syntex Institute for Molecular Biology in the Stanford Industrial Park. With Terman allowing him to continue teaching, Djerassi agreed to run the institute and signed up Joshua Lederberg, a Stanford professor of microbiology and 1958 Nobel Prize winner, to serve as the institute's advisory research director. Lederberg laid out the institute's research agenda.

Three years later, when Syntex decided to enter the U.S. market under its own name, Djerassi persuaded the company to set up in Palo Alto, far from the pharmaceutical corridor of northern New Jersey; as the only drug house west of the Mississippi, Djerassi argued, it would have the pick of the best researchers in the western United States. Like so much else, Syntex's style of corporate development broke with traditional pharmaceutical practices. Just as Ogden had spun off Syntex to shareholders, so Syntex then spun off a series of new science-based companies, including Alza (drug delivery systems) and Zeocon (an attempt at steroidal-based insect control run by Djerassi himself). A third unit, Synvar, began as a joint venture in superconducting polymers between Syntex and Varian, an electronics company founded by former Terman students and based in the park, but quickly evolved into an innovative diagnostics unit. In 1977 Syntex renamed the unit Syva and made it a full subsidiary of Syntex. The firm also set up subsidiaries in animal health and dental products.

The argument for diversification can always be put in two ways. On one hand, the diversifying firm can argue—as so many conglomerates did—that it is going into a new business to find more productive use for its capital. Or it can claim that it is taking technological skills from one product area and using them in another. Syntex made both arguments. But diversification in a technologically driven industry as potentially profitable as pharmaceuticals always had another subtext: that the engines of drug R&D were not powerful enough, or predictable enough, to generate enough profitable new products; therefore, why not pour capital into less profitable, if steadier, businesses, like animal health, cosmetics, or over-the-counter products? Although the equity markets embraced the logic of diversification in the 1960s and 1970s, they can hardly be accused of bullying the drug companies to put their capital to nondrug uses: most drug companies rarely dealt with the equity markets, and besides, the markets tended to value drug R&D very highly, recognizing the powerful effect on profits of a single new blockbuster.

In pharmaceuticals, the decision to diversify tended to come from top managers insecure about the returns of drug discovery (the exception that circumstantially proves the rule here is that the most successful drug houses, Merck and Glaxor, diversified the least). Syntex was no different than the rest. Despite its steroidal triumphs, Syntex did not get much further than any of the other drug companies in rationalizing its drug discovery program. Innovation had slowed, and by the 1980s Syntex had one of the poorer records among the big drug companies for producing new drugs, despite continuing to pour heavy investment into R&D.[2]

Notwithstanding those failures—and declining R&D productivity plagued the entire pharmaceutical world in the 1980s—few would question Djerassi's choice of molecular biology as the area where the opportunities for rational drug discovery appeared to be the greatest, even to traditionalist drug companies.[3] Molecular biology made the most persuasive case for providing the scaffolding within which a rational discovery program could be developed. And out of molecular biology would emerge the biotechnology industry, a collection of new, biologically oriented companies that declared in the early 1980s that the tyranny of size and financial power had been broken and that their particular strengths, speed, and biological insight would triumph over the financial power and reach of the drug elite. Fueling that triumph, of course, was the widely held belief, one that seeped even into the pharmaceutical establishment, that the secrets of a rational discovery process were about to be revealed by molecular biology.

What is this discipline called molecular biology, and what are its links to the industry we now call biotechnology? The term *molecular biology* was first coined by Warren Weaver, the director of the Rockefeller Foundation's Division of Natural Sciences. In the early 1930s, Weaver began to funnel money toward projects that used chemistry and physics to study biological questions (the notion that there was a molecular basis for biological phenomenon had appeared intermittently in biology, but it was not a widely accepted idea). Weaver directed funding particularly at efforts to untangle the molecular basis of inheritance and at attempts to understand the structure of large biological macromolecules, such as proteins, through X-ray diffraction techniques. In 1938 Weaver hammered home the case for this new discipline in the foundation's annual report: "Among the studies to which the Foundation is giving support is a series in a relatively new field, which may be called molecular biology, in which delicate modern techniques are being used to study even more minute details of certain life processes."[4]

Molecular biology had a complex, if distinguished, pedigree. The central biological insight of the nineteenth century, Darwin's theory of evolution, had driven biologists to the building blocks of the cells in their search for a mechanism that determined inheritance and governed reproduction. For much of the twentieth century, the investigation of the material that filled the cell nucleus, the so-called nucleic acids, was strictly a chemistry problem, and one that initially provided very wrong answers, generated in part by the still-low technical level of the science. For decades, the harsh methods of organic analysis inadvertently broke down the long chain molecule now known as deoxyribonucleic acid (DNA), convincing chemists that it could never play a major role in inheritance.[5] In the 1930s and 1940s, with the development of polymer chemistry, biochemists began to develop better analytical methods (Weaver's "delicate modern techniques"), such as ultracentrifuges. Only then did DNA's immense length, far larger than any known protein, become apparent.

By the 1940s, some physicists, now self-confidently in the van of scientific progress, headed toward biological questions. Structural chemistry, as pioneered by the likes of California Institute of Technology chemist Linus Pauling, owed an enormous amount to quantum physics. During and after the war, a number of the quantum physicists, such as Neils Bohr in his 1932 lecture "Light and Life" and Erwin Schrödinger in his 1944 lecture "What Is Life?," either commented on biology or actually plunged into it full-time, like Max Delbrück and Leo Szilard. Delbrück in particular assumed a central role as a sort of biological theoretician (the very idea of overarching biological theory was scandalous to traditional biological empiricists) and a leader of an American movement known as the phage group, which used the bacterial virus as a model to study molecular genetics.

The physical link to inheritance was finally unveiled in 1944 by a Rockefeller University team led by biochemist Oswald Avery, who was investigating what he called "the transforming principle" in pneumonia bacteria. That material turned out to be DNA. Still, no one knew what DNA was or how it worked. Those large questions would not begin to be answered until 1953, when a young American biochemist and junior member of Delbrück's phage group, James Watson, and a slightly older English physicist turned biologist, Francis Crick, jointly elucidated its structure in what is now one of the most famous episodes in modern science. The structure of DNA turned out to be functionally elegant: a double helix containing four paired constituent amino acids—adenine with guanine, ctyosine with thymine—the order of which then determined the genetic code.

The elucidation of DNA was one of those scientific discoveries that seem tuned to the zeitgeist. The notion of a genetic code evoked the code-breaking machines of World War II, the contemporaneous development of cybernetic theory by mathematicians John von Neumann and Norbert Weiner, and the coming development of advanced literary theory, with its codes, subtexts, and deep structures. Molecular biology was also interdisciplinary, part chemistry, part physics, part biology. In more general terms, the apparently simple relationship between structure and function in DNA raised the hopes again that biology could be as predictive or "rational" as mathematically derived physics—more tangibly, that molecular structures could be designed that could perform specific biological functions. This prospect was made more tantalizing when DNA was portrayed as another controlling metamolecule, like the steroids, the fashionable analogy being to a biological von Neumann computer. Program the computer differently, and you would get different results.

It is easy to get swept up in the grandeur of an ascending paradigm. But there is no doubt that the elucidation of the double helix initiated the long, expansionary program of molecular biology that changed the face of biology. "There has hardly been a more decisive breakthrough in the whole history of biology," writes Harvard's Ernst Mayr in his history of biological thought.[6] Horace Freeland Judson compares the rise of molecular biology (and quantum physics) to that of a dynastic ascension of some early modern European state:

> Revolution takes place within a frame of comparatively unyielding continuity. . . . Molecular biology is no single province, marked off by natural boundaries from the rest of the realm. It is, rather, an intellectual transformation—indeed, a new conceptual dynasty—arisen within the realm. As a dynasty, molecular biology is by no means identical, any longer, with its ancestral heartland, the physical chemistry of the gene-stuff. It has a history now, and, some claim, a culture of sort, and others have charged, an ideology. It is expansionist. . . . Molecular biology is a discipline, a level of analysis, a kit of tools—which is to say, it is unified by style as much as by content. The style is unmistakeable. The style is bold; it is simplifying; it is unsparing; often it is extremely competitive. The style is also, sometimes, subtle and sophisticated.[7]

Dynasty. Transformation. This is stirring stuff. No wonder molecular biologists developed a reputation for arrogance. But what exactly *is* molecular biology? On one hand, it is the exploration of biology at the molecular level—a definition so broad as to be almost useless. More specifically, it is the painstaking unraveling of the way in which DNA and the various RNAs (ribonucleic acids) mastermind reproduction. Molecular biology is also the analysis of the structures of proteins. While some observers argued that the discovery of the double helix was less a revolution than the culmination of decades of biochemistry, it certainly created the spark that altered the structure of biology as an academic discipline.[8]

Most molecular biologists did not worry unduly about definitions and distinctions. In its youthful growth phase, through the 1960s, molecular biology defined itself much as quantum physics had: as a discipline exploring the most fundamental order of biological reality. Judson quotes Crick: "Molecular biology is whatever interests molecular biologists," a statement as subjective and arrogant as that of any physicist or, indeed, any entrepreneur.

## The Advent of Biomedicine

By the mid-1960s, a formidable gulf still separated biology from drug discovery. That gap was embodied in the way the biomedical funding apparatus had evolved since the war. In the drug industry, large, stable cash flows from a small cohort of products financed the complex organizations needed to support industrial programs of mass screening and FDA-mandated clinical testing programs. Likewise, swelling biomedical funding from the federal government, for the most part insulated from the markets, appeared to be an effective way to support the kind of basic research, particularly molecular biology, taking place in academia. With the occasional exception of a project like the Syntex Institute for Molecular Biology, those two financial streams—government funding into biology, mostly in academia, corporate funding into drug discovery— remained separated. Few in the government or industry worried that there was only nominal contact between the two spheres. Biologists, in particular, were proud of their purity from industrial ties.

That sharp division began to blur in the early 1970s as the gap between medicine and biology narrowed (the very use of the term *biomedicine* is a tribute to that convergence).[9] The factors driving this process were quite involved, reflecting the social complexity of the modern scientific enterprise: from the government (the war on cancer), academia (the development of genetic engineering and the rise of the immunotherapies), and the economy (the inflation of the 1970s, the deregulation of Wall Street, various tax reforms). Each of these factors altered, however subtly, prevailing ways of thinking. For biotechnology to be born as an industry, not one, but two traditional mind-sets had to be renovated—one in academia, the other on Wall Street and industry. In academia, many biologists had to become convinced that the wall separating biology and industry served no particular purpose and had to come down. And on Wall Street, investors had to conclude that biotechnology was a sort of biological proxy to microelectronics and computers. The final step on the road to a biotechnology industry resulted when these two outlooks merged: business as science, science as business.

Academic molecular biology had been an expansionary field—competitors for biomedical funding often denounced it as hegemonic—since the discovery of the double helix. Just as it did in solid-state physics, the growth of federal funding in the 1950s expanded the academic base of biology, producing by the late 1960s the first population pressures— though, unlike in physics, there were few commercial outlets. Those pressures were at first relieved, then renewed, with President Nixon's 1971 announcement of a war on cancer, an attempt to direct science toward a government goal. By then, molecular biology was no longer just an intimate community of teachers and students. There were now lab chiefs and major investigators, technicians and postdocs, the tenured and the untenured. As it matured, the community itself split into specialized factions—virologists, immunotherapists, the DNA and RNA crowd, the biochemists, the cell development folks, and, in time, the cloners and sequencers. It also split by age and class, between those established and the far larger numbers seeking their funded niche, between those at the center and those on the periphery. At the top of this scheme sat the laboratory chief, a powerful academic entrepreneur, a veritable scientific superstar if he had a Nobel, who organized a middle management of section chiefs and investigators that in turn rested upon a working class of postdocs, grad students, and technicians.

These pressures created fertile ground for a massive shift into commerce. The actual catalyst for much of this shift, and for the birth of biotechnology, was a series of technical breakthroughs that culminated in Herbert Boyer's and Stanley Cohen's first successful recombinant experiments in 1973, which successfully implanted DNA from a frog into a common colon microbe, *Escherichia coli*. Genetic recombination created a tool that offered the possibility—one that seemed to be so very close to becoming reality—that molecular biology, and thus biological inheritance and design, could be altered and manipulated like any other engineering discipline. It was a potent version of the rational dream, held up as evidence that the gap between biology and medicine was closing—an eventuality molecular biologists had preached about for years—and that revolutionary developments would quickly follow. That possibility, in turn, unleashed two seemingly contradictory movements. First was a flurry of doubts, soul searching, and fractious controversy over moral and social issues; then came a pell-mell stampede into commerce. Beneath all the turmoil ran a deeper theme: academic molecular biology's attempt in the 1970s to define and then alter its view of itself and its place in the world. Was molecular biology a sort of Platonic republic of science, self-contained, self-regulating, a model of free inquiry and open communication? Could it continue to strive for that ideal as the science matured—as it became more powerful, more valuable? And if that was the case, what were the ties—and responsibilities—to a greater nonscientific, democratic public?

In retrospect, the events of the 1970s in molecular biology have a naive quality. It is not that the fears that motivated these events have been eliminated; rather, it's that some of the issues of that time have been resolved and others seem merely irrelevant. Biology, even in academia, no longer belongs to the biologists alone. Like the atomic scientists immediately after the war, molecular biologists in the 1970s believed that they could control the forces they had unleashed and that molecular biology could operate as a large, powerful community, as it had when it was a small, obscure sect.

## The Changing Role of Biology

In the early 1970s, tensions among the various factions and classes in the biological community were only just beginning to be felt. Then in 1973 came the successful Boyer and Cohen recombination. In June the experiment was all the talk at the Gordon Conference on Nucleic Acids, a prestigious annual forum for molecular biologists held in New Hampshire. Fearful of the cancer-causing effects that could occur if an altered virus escaped the lab, the conference participants wrote to the president of the National Academy of Sciences expressing their "grave concern. . . . These experiments offer exciting and interesting potential both for advancing knowledge of fundamental biological-processes and for alleviation of human health problems. Certain such hybrid molecules can present a health hazard for the laboratory workers and the public."

The controversy, so far confined to scientific circles, was already racing ahead.[10] No one wanted to unleash some Andromeda strain upon a helpless population (Michael Crichton's novel *The Andromeda Strain* was published in 1969; the movie came out in 1971). On the other hand, any attempt to freeze research ran against even more basic drives: curiosity, free inquiry, the progress of science, continuing funding, fame, power. The National Academy of Sciences formed a committee on biohazards headed by Paul Berg, a molecular biologist from Stanford, which included Watson, Boyer, and Cohen. The Berg report came out in mid-1974 urging certain safety restrictions when experimenting with recombinant DNA products and calling for a moratorium on certain experiments until more was known. This was an unparalleled exercise—a group of scientists unilaterally deciding that a broad and exciting area of experimentation should not be pursued. In March 1975, almost two years after the original Boyer-Cohen experiment, 140 of the most prominent molecular biologists in the world gathered at a lush California state park called Asilomar on the Monterey peninsula to decide what to do next.

Asilomar marked the height of molecular biology's idealist phase.[11] The meeting participants hardly agreed on everything, but the urge toward self-regulation was very strong. On that consensus, the meeting slipped past knottier contradictions. This was, after all, a self-selected (and international) elite; and it was an elite that was supported almost entirely by public funding. The proceedings themselves were run in the sort of labored openness that characterized the student takeovers of the universities just a few years earlier, with a set of issues that the profession was fundamentally in accord with. The only jarring note came when a lawyer who had been invited to speak warned of liability questions; the notion that there could be possible legal penalties unsettled the group. Watson, the spiritual leader of American molecular biology, summed up the feeling a few years later when the backlash had already begun, in a book on the biohazards controversy he wrote with John Tooze:

> Although some fringe groups (such as Science for the People) thought this was a matter to be debated and decided by all and sundry; it was never the intention of those who might be called the Molecular Biology Establishment to take this issue to the general public to decide. The matter was not only too technical but in a way too fuzzy for responsibility to be shared easily with outsiders. We did not want our work to be blocked by overconfident lawyers, much less by self-appointed bioethicists with no inherent knowledge of, or interest in, our work. Their decisions could only be arbitrary. Given that there were no definite facts on which to base "danger" signals, we might find ourselves at the mercy of Luddites who did not want to take a chance on any form of change.[12]

To avoid the Luddites and lawyers, the conference came up with a complex, rigid set of guidelines, which, in due course, was approved by the National Institutes of Health and monitored by a committee called the Recombinant DNA Advisory Committee, or RAC. With these controls in place, the moratorium ended and experimentation began again—though constrained by RAC rules and by the fact that safer laboratories sometimes meant higher costs. RAC had no legal powers, and the moratorium did not extend beyond federally funded research. Asilomar, for all its idealism, underscored the reality that molecular biology was increasingly a game to be played by the well funded, and in academia that meant those who held power in the traditional peer review system. And even as the establishment demonstrated a remarkable coherence, at least at the top, its power was eroding—particularly, though not exclusively, at the bottom. Small companies—insignificant at the time— were forming on its margins (they did not have to abide by the moratorium nor by the RAC rules); and for all its idealism, Asilomar triggered a wave of intense academic competition for money and brains.

For all their attempts to do so, the establishment could not insulate these issues from the larger public. For one, the press found biology a source of sensational stories. Also, nonscientific critics of experimentation and testing, such as Jeremy Rifkin (whose first broadside, with Ted Howard, *Who Shall Play God?* appeared in 1977), found that these issues were a crusader's dream.[13] In 1977 Rifkin compared recombinant work to Nazi experimentation on unwilling victims, enraging scientists. Academia itself was deeply divided, despite the efforts of Watson's establishment to paper over the differences. For those who did not belong to that establishment, or even for those who did but found themselves in fundamental disagreement, the public, through the media and through politicians, beckoned as another avenue of recourse, first in social and academic controversies and then, inevitably, as a source of funding.

The cockpit of the social and political debate was Cambridge, Massachusetts. The first controversy arose over the construction of a new P-3 biological laboratory at Harvard in 1976. Harvard scientists, including Watson, Walter Gilbert, and Mark Ptashne, argued that they needed a new laboratory if they were going to keep pace with developments occurring on the West Coast. At the time, Gilbert was racing to clone and sequence the insulin gene—a contest also involving a new San Francisco company called Genentech working with a group at the University of California at San Francisco. Although Harvard agreed that a new lab should be built, not everyone on campus, or in Cambridge, was satisfied. The dissenters argued that such a laboratory should be isolated from large numbers of students and faculty; it should not be, as it was planned, on the third floor of an already crowded facility. One of the leaders of the group was Ruth Hubbard, a tenured Harvard biologist. The critics finally decided to go outside the university community. "The sin," Hubbard said, "for which there is no forgiveness."[14]

The result was a long June 8, 1976, article in an alternative paper, the *Boston Phoenix*, which laid out the dangers of recombinant DNA work off Harvard Square. "This is a biohazard," the paper quoted Hubbard. "It's not enough to stop doing the experiments. Once it's out it's impossible to shut down. It's worse than radiation." The article caught the attention of Cambridge mayor Alfred Vellucci, who then scheduled public hearings. The debate, before television cameras and the assembled press, quickly focused on the luridly fantastical side of genetic experimentation, "the monsters in the sewers" problem. While Vellucci's confused statements were sensational—and attention getting—some major scientific names, most associated with the Science for the People movement, also testified against the facility: Hubbard; her husband, Nobel laureate George Wald; Jonathan King of MIT; and Jonathan Beckwith of Harvard Medical School. There was no unanimity in the scientific ranks. And on an issue of such importance, the critics were willing to seek support from a nonscientific public.

After the sensational aspects of the case receded—following a three-month moratorium at Harvard and the establishment of a joint Harvard-Cambridge safety committee—the university proceeded to build its P-3 laboratory. But something was lost. The theoretical democracy of the scientific community, erected upon merit and peer review, was perceived to have broken down, undermining the mechanism of self-regulation. Moreover, the behavior of Vellucci confirmed the biological establishment's worst fears, just as it was growing more confident about biohazards and more aggressive scientifically. In 1977 and 1978, RAC progressively loosened its guidelines and, six months later, the suggestion was made to drop them entirely. In September 1979, the Asilomar guidelines were all but abandoned, but not before a haphazard jumble of local and state regulations took their place. Many, however, recognized a new reality: once consulted, the public could not be easily ignored.

A second set of disputes added fuel to the biohazards fire. At issue: How far should universities go in soliciting commercial support? What constituted a conflict of interest? What should be the relationship between academic science and commerce? In California, Herbert Boyer was fiercely attacked for his ties to the newly formed Genentech—he was cofounder, scientific adviser, and a big shareholder—though he never left his academic post at the University of California, San Francisco; there were even rumbles that he had lost a Nobel because of the controversy. Similar disputes arose at other major research campuses, particularly over patents.[15] But the most acrimonious, and most revealing, dispute took place again in Cambridge. In the early 1970s Edwin "Jack" Whitehead, cofounder and president of Technicon, a pioneering medical diagnostics company, began to look for a way to make a major philanthropic contribution by setting up a research institute in alliance with a university. He approached Stanford, Caltech, and Harvard and actually began negotiating with Duke in 1974. But nothing came of those contacts.[16]

In the mid-1970s Whitehead met David Baltimore of MIT, a molecular biologist and Nobel laureate, and persuaded him to head up the center. Baltimore convinced Whitehead in turn of the merits of MIT, with its long tradition of mutually beneficial corporate–university ties. In 1981 a formal proposal was made and the battle joined. Whitehead was willing to put up $120 million, the largest single research gift in fifty years, to retain control over faculty appointments who would also be a part of the larger MIT faculty. Such control was necessary, the pro-Whitehead forces argued, to attract and retain the necessary high level of talent. Those opposed to the institute argued that Whitehead appointees would be a large enough bloc shift the balance of power on the MIT biology faculty and open it up to commercial pressures.

The surfacing of these issues (and, in nearly every case, the eventual resolution in favor of outside funding or outside commercial ties) further eroded the traditional biological ideal and the sanctity of academic research. If the biohazards question pivoted on medical debating points, the conflict-of-interest issue hinged explicitly on financial ones.[17] The universities needed the money not only to support current facilities but to build new ones—that is, to compete in the research game. Increasingly in the 1980s they couched their arguments in economic terms: the funds were necessary so that America could compete. Even during the Harvard laboratory debate, the university complained that it was losing talent to better-equipped institutions, particularly in California. Eventually, the bogeyman would be Japan.

The universities felt competitive pressures for tangible reasons. Federal support had begun to flatten in the late 1960s.[18] First Vietnam and the Great Society (particularly Medicare and Medicaid), then general economic distress (recession, double-digit inflation, lagging productivity growth), absorbed funds and slowed the growth of the federal biomedical research juggernaut in real terms. Between 1966 and 1982, biomedical spending growth slowed to 3 percent annually, despite the inauguration of the war on cancer in the early 1970s, considerably less than the 16 percent growth between 1950 and 1965. Some growth was better than no growth, of course, but 3 percent was not enough to support the growing population of molecular biologists. In particular, the NIH, the major backer of academic biology, found it more and more difficult to invest in younger talent. Not only did the average size of NIH grants decline in the 1970s, but NIH went from funding two-thirds of all approved competing projects in the early 1970s to funding only one-third a decade later. Meanwhile, "indirect costs" for university overhead and plant went from about 20 percent of all grants to 33 percent. The universities also sensed the first pressures from small start-ups. Individual faculty members, a few quite prominent, began to consult and even to leave academia to launch commercial enterprises. Baltimore argued that the Whitehead bequest was necessary for MIT to avoid more, increasingly intrusive, commercial ties in the future.

Whitehead and Baltimore won. The Whitehead Institute for Biomedical Research was founded and housed in a handsome structure on the MIT campus. Whitehead appointees became part of the MIT faculty, and the institute went on to produce first-rate basic science. Little is heard today about the conflict-of-issue question, which is a sign of how potent, on all levels, the competitiveness issue has become. The visceral fears that outside interests would distort the search for truth and would taint academic freedom have all but disappeared. It is not that those fears and conflicts do not still have meaning; rather, in a world where there is constant interchange of manpower and money between academia and commerce, few seem to care much publicly anymore.

While those debates were agitating campuses, interferon, a naturally produced protein of uncertain potential, was driving the issues into a new region: peer review. Interferon was a mysterious antiviral protein that had been discovered in 1957. After several decades of sporadic research, little was known about it except that, like the steroids, it was both rare and very expensive to isolate. Nonetheless, in the mid-1970s, Mathilde Krim, a politically well-connected former Memorial Sloan-Kettering researcher who had helped write a congressional committee report for the war on cancer legislation, began to promote interferon as an anticancer agent that had been ignored by a bureaucratic establishment intent on the status quo. As the 1970s wore on, and the war on cancer proceeded, the demand for a true cancer cure grew. Krim began to organize and promote the still-obscure protein and market it, mostly through the media, to a new public constituency.[19]

Interferon was the catalyst that triggered the first wave of enthusiasm for biotechnology on Wall Street in the early 1980s. It was biotechnology's first great "concept" and attracted the first wave of public investors. Perhaps more significant is that interferon was the excuse for abandoning traditional means for new ends. Krim sought to change the underlying rules by which biomedical research was conducted. She bypassed the conservative judiciary of peer review; like scientific activists in other spheres, she actively nurtured a nonscientific constituency that, in turn, could apply pressure to the funding establishment. She, in effect, opened the door to public investors and invited them to draw up their own research agenda. After interferon, the role of the public—particularly the segment known as Wall Street—became a permanent part of the agenda-setting system in biology.

Funding pressures within academia also fed the interferon boom and contributed to the launch of biotechnology. The advent of immunotherapy, another metasystem susceptible to public acclaim and distortion, provided a theoretical justification for promotion. The Reagan attack on centralized authority, from regulatory agencies to social welfare agencies, was also attracting support. The laetrile fad that erupted publicly in the late 1970s revealed a deep frustration and profound distrust of established institutions like the National Cancer Institute and the FDA. And, of course, between Asilomar in 1976 and interferon's great boom in the early 1980s, the community of molecular biology had changed. The debates, the battles, the scramble for funding, the continuing scientific developments—all these chipped away at the strict separation of biology and industry. There are various ironies to this transition. Many within the establishment who had once fervently believed, with Watson, that science should retain control, now rushed to raise money on Wall Street, effectively ceding control. A number of the elite at the top of the peer review pyramid were the first, in a commercial context, to announce breakthroughs before any confirmation. And many of those quickest in the biohazards debate to seek alliances outside the community—to share the science—now were the quickest to warn of the dangers of commerce.

## A Weakness for Miracles

Wall Street, like molecular biology, also had to alter its modus operandi to accommodate new forces in the 1970s. In particular, Wall Street had to accept the notion that biotechnology, for all its complexity, resembled microelectronics and computers—biotechnology was just newer and thus potentially faster growing. And to do so meant funding companies far earlier in their life cycles than public markets had ever done on such a scale before. In short, the evolution of Wall Street in the 1970s involved two movements: the strengthening belief that biotechnology constituted a technological revolution, and the explosion of what might be called *public* venture capital.

By the 1970s, Wall Street had had considerable experience with small biomedical companies. But these tended to be medical device, diagnostic, and hospital supply companies that required relatively little capital to expand already existing product lines; in many ways they were no different from machine-tool firms, makers of auto parts or scientific instruments. Although these companies were not very exciting, occasionally they were quite profitable (as Technicon or Beckman Instruments were), particularly if they were eventually acquired by a larger corporation (as Smith Kline & French acquired Beckman and Johnson & Johnson bought Technicon). Into this mix came brash, imaginative Syntex, with its alluring blend of creative science and business. Syntex growth had slowed by the 1970s, but it left behind an aura of possibilities.

Syntex had fed Wall Street's weakness for miracles. In the economically grim 1970s, the technological growth stocks of the 1960s, such as IBM, Xerox, Polaroid, DEC, and Syntex, took on the warm glow of legend. Wall Street itself was rebuilding from the rubble of the post-1960s crash, consumed by wrenching changes such as the deregulation of brokerage commissions in 1975, and it was transforming itself into a harder, faster, more market-oriented arena, more a meritocracy than a club. Like molecular biology, Wall Street found that it had to abandon some of its less utilitarian ideals in the 1970s—notably, the old reliance (in spirit if not always in fact) on relationships. Although many individual investors were driven out of the equity markets after the bear market of the early 1970s—many of them ended up in mutual funds, which of course often invested back in the market—institutional investors continued to grow and dominate the new Wall Street. Automation was also coming fast and forcing Wall Street to restructure and consolidate. In 1971 the National Association of Securities Dealers created an over-the-counter market of linked terminals powered by mainframe computers in Trumball, Connecticut. NASDAQ fit somewhere between the more established New York and American exchanges and the shadowy world of penny stocks, and it was driven by the growing interest of investors in smaller companies that, if they took off, provided blockbuster, venture capital–type returns.

Biomania began, as the first wave of biotech companies went public, followed by chip, computer, and computer peripheral companies. Shares of Genentech, a biotechnology company from San Francisco, exploded onto the public markets on October 14, 1980, rocketing from an offering price of 35 to 89 in the first twenty minutes of trading. As the fever rose, quality declined. On the margins, a handful of smaller brokerage houses sent large numbers of young firms, most having little contact with long-term venture capital but boasting exciting "concepts," into the markets. Investors eager to get in earlier in this hot game, prone to conceptual siren songs, snapped them up. Thus arose the phenomenon of public venture capital, a trend that would attain its apotheosis with a flood of biotechnology offerings in 1981 and 1983. While some biotechnology start-ups would go public earlier in their life cycles than others—a few mere months after incorporation—no major biotechnology company sold equity to the public before it had any hope of actually selling products. This was in sharp contrast to the chip and computer start-ups of the same period, many of which had been nurtured during the latter 1970s. In a few cases, the institutional money manager, operating through anonymous, flighty public markets, served as the venture capitalist for small companies—a situation fraught with risk.

Georges Doriot's sense of loyalty never seemed more antiquated. The markets for biotechnology companies, which attempted to predict the course of a complex, extremely murky biology, were clearly "less perfect" than markets focused on more established firms or on companies exploiting more rational technologies with shorter development cycles, notably microelectronics. Lacking other benchmarks, biotechnology investors embraced the reality of Drucker's knowledge worker, and a cult of personality developed. What mattered to investors in biotechnology in the early 1980s was not bricks and mortar, nor patents and contracts, but the intellectual capital that resided in a scientific founder, a research director, or an advisory board studded with Nobel laureates. How to measure that intellectual capital? Reputation and credentials, most often derived from success in academic research, meant everything.

Charisma generally was in demand in the feverish markets of the 1980s, just as it had been in the 1960s go-go market. In computers, the charismatic figure would be embodied in Steven Jobs, the cofounder and house visionary of Apple Computer; later, in junk financing, Michael Milken served as the man in the vanguard, the only man in a market of imperfect data who had full knowledge and control. Biotechnology threw up a number of different personalities that represented the point where academia and business crossed, producing a new generation of scientific entrepreneurs. The pendulum by now had fully swung. American industry was viewed as slow moving, dull, and uncompetitive (and corporate raiders were just gearing up to do something about that). The antidote was the entrepreneurial hero, particularly the kind that wielded a technology that could slay, like Luke Skywalker, the dark forces of bureaucracy. The entrepreneur radiated, figuratively and often literally, youth; and the entrepreneurial start-up, with its creativity and flexibility, stood in sharp contrast to the arthritic stiffness of larger, older organizations. The entrepreneur was thus portrayed by the media and by commentators such as George Gilder as a sort of revolutionary, able to leap the abyss of high risk, break old rules, release pent-up energies, produce scientific miracles, and restore the economy as well. "Entrepreneurs everywhere ignored the suave voice of expertise," wrote Gilder. "Confronting the perennial powers of human life, the scientific odds against human triumph, the rationalistic counsels of despair, the entrepreneur finds a higher source of hope than reason, a deeper will of faith than science, a further reach of charity than welfare."[20]

That message proved seductive both on Wall Street and in certain precincts of academic molecular biology. Wall Street wanted to believe in a new wave of technological growth stocks; molecular biology wanted to believe that this new kind of company—academic biology, so many mistakenly thought, dressed up with stock options—solved the problem of shabbily genteel salaries, intense competition, and tightening resources in academia. Each side viewed the other in simpleminded ways. Wall Street imagined that academic biologists could harness unlimited technological powers, and thus business experience meant little; the academics saw Wall Streeters as either a bunch of easily manipulated brokers or as wily and greedy speculators. In the 1970s, the biological community had argued its way toward an accommodation with the public; turning to public investors for financing was going a step further. From the beginning, it was clear that the public was split on the subject of advanced biology: some embraced the horror stories of Rifkin, while others swallowed whole the tales of miracles and money. Wall Street reflected those views, either driving biotech stocks up or selling them off wholesale. Like the images of the entrepreneur as a divine hero and the bureaucrat as a villain, both views of biotechnology were overly broad. Biotechnology was new, but it was not *yet* revolutionary; it was powerful, but not transformative. Beneath the emotions stirred up by biotechnology, it was a business like any other.

## Genentech: The Perils of Integration

A few small, venture capital–financed, biologically oriented companies had appeared prior to 1974—most notably, Cetus, based in Emeryville, California, near San Francisco, and Collaborative Research, in Lexington, Massachusetts, a few miles west of Cambridge (both were physically located, significantly enough, near academic and venture capital hotbeds).

The most important start-up in biotechnology came in 1976, when Genentech was formed after a meeting in a San Francisco tavern. Genentech contained all the various ingredients for biotechnological credibility: a well-known academic scientist (Boyer); a venture capitalist turned entrepreneur (Robert Swanson); a source of professional venture capital (Eugene Kleiner's San Francisco venture firm, now called Kleiner Perkins Caulfield & Byers, where Swanson had been a junior partner); and a powerful technology (recombinant DNA).

Genentech would trigger the first bout of biomania on Wall Street.[21] And in the years ahead, the company would set many of the major commercial and technological firsts in biotechnology, from limited partnerships, to aggressive patent litigation, to the first recombinant products. Genentech also proved the most dramatic victim of Wall Street's moods: highly valued throughout the 1980s, Genentech found itself cast out by investors when overheated expectations were not met. In 1990 Genentech agreed to a partial merger with Roche, the big Swiss drug company.

Genentech symbolized the tumultuous first decade of biotechnology, not only for its indisputable technical and commercial achievements but for its eventual failure to generate self-sustaining profits, to achieve what economist W. W. Rostow (speaking of developing economies) once called "take-off." What was the difference between a Texas Instruments in the 1950s and 1960s and Genentech in the 1980s? Many on Wall Street in the early 1980s would have said that there was no difference of note. Biotechnology was the next wave, the technological successor to the semiconductor industry, a field dominated by fast, agile competitors able to ride extremely rapid, transformational change. The fact that some of the better biotechnology companies were funded by famous semiconductor venture capitalists drove that point home. "Just as the high technologies of the semiconductor and computer industries created new markets and new ways of doing things, the fundamental new technology of genetic engineering is rapidly developing into an important new business that will have an impact on science as well as business," began Genentech's first annual report as a public company in 1981. The company then quoted the *New England Journal of Medicine:*

> Never has there been a comparable period in the growth in the knowledge of living things. The current evolution in life sciences is not a simple, linear projection of the growth curve in knowledge for the past century. A striking shift in that growth has occurred, amounting to a geometric progression of information. . . . We are on the threshold of some unusual transformation in health practices, agriculture and industry.[22]

This transformation science—molecular biology—was widely viewed, in academia and Wall Street, as capable of remaking the commercial world controlled by the established drug companies. Biotechnology was an industry that would grow extremely quickly by producing new and better replacement products for the limited ethical drugs of the past. It was a Schumpeterian industry that would toss up fully integrated corporations to replace the hierarchical, slow-moving bureaucracies of the past. In 1982 the chief executive of Hybritech, another Kleiner Perkins–backed biotech company, told Wall Street analysts, "No industry remains static in the face of major technological breakthroughs, and what's happening today parallels what Merck, Lilly and Pfizer achieved several decades ago when antibiotics were discovered. . . . Today, there are skeptics that say about Hybritech: 'It's pie-in-the-sky technology that's years away'; 'Hybritech's a research boutique, like most biotechnology companies'; 'Big companies will take over'; and 'They're still a venture capital deal in the early stages of putting a company together.' My objective is to destroy those preconceptions."[23]

What would a group of small companies have to do to achieve those goals? They would have to meet three requirements. First, enough capital would have to be available to support a threshold level of R&D. Although no biotechnology company came close to outspending even one of the smaller elite drug companies, biotechnological R&D was still a capital-intensive effort, far more so than R&D in the early days of transistors. Thus, a biotechnology industry, like the start-up custom chip industry of the 1980s, would never have arisen without the presence of active venture capital and generous public equity markets. Because so much capital was invested in the 1980s, literally hundreds of companies claiming to use biotechnological methods were launched. The downside of this financial largesse was that it spawned a feeding frenzy similar to the funding of disc drive makers and custom chip companies in Silicon Valley, around similar products attractive to investors (the interferons, the interleukins, various antibody therapies), which thinned out expertise, provided a lawyer's heaven of patent suits, arguably misallocated capital to worthless ends, and generally confused already baffled investors.

Second, simply generating capital was necessary but not adequate. The ascendant industry would have to use the available capital more *productively* than its established competition. Thus, in one sense, the war would be fought on the field of innovational productivity.

Finally, the industry would have to commercialize its products more *quickly* than the established firms (that is, the time between when the company began spending on research and when the resulting product actually generates sales would have to be shorter). Compressing the development cycle would reduce dependence on capital markets and cut costs; time, in this case, is analogous to money. Biotech companies had to move dramatically more quickly or more cheaply, or both, than their established competition. The early transistor makers such as Transitron and Texas Instruments routinely beat the vacuum tube makers to the market with new transistors, gradually adding market share. In biotechnology, a similar process would have to be sustained over a period of years for the industry to achieve full integration, takeoff, and victory.

These factors are, of course, interrelated. The acceleration of new products required to generate takeoff depended not only on how quickly the established firms developed new products but on the availability and cost of capital. Texas Instruments was dependent on the capital markets only for short periods of time and for small amounts; capital raising was quickly followed by new product introductions, and of course the company had a stable source of earnings in its other businesses. Most biotechnology companies were quite different than that. They raised capital to perform fundamental research years before commercialization began and profits came. Few were able to generate financing to fund an entire product cycle at once, and so they continually returned to Wall Street, begging bowl at ready, often to discover that investors had turned against high-risk ventures just when they needed the help the most. Failing to raise money by selling equity, these firms were forced to auction off assets (from product licenses to the company itself, depending on their plight), breaking down integration.[24] The result was that the established drug companies received the fruits of the first decade of biotechnology by simply waiting for the money to run dry. New products fell into their laps. Biotechnology may have been more innovative and productive than the established drug companies, but not enough to wipe out the advantage of financial power and organization.

Technological development in biotechnology also lacked the predictability of transistors. The development cycle at Texas Instruments was very short, very cheap, and very predictable—or rational—compared with that at any biotechnology company. The biotechnological development cycle was fraught with unexpected, and often unwanted, surprises (just like in the drug industry); many of the most heavily publicized products were the ones that had had the most tortuous, high-risk path to the market. Few emerged unscathed. In short, while biotechnology dealt with the same tradeoff between time and money as Texas Instruments and Transitron in the 1950s, the results were radically different because the underlying sciences were at such different stages of development. It was the rare venture capitalist or investor that recognized those differences.

Genentech, the pacesetter and pioneer, was representative of this complex interplay between innovation and finance. By 1986, Genentech had raised over $75 million in two public equity offerings and another $106 million in three limited partnerships that were designed to fund the development of specific products. Like most biotechnology companies, Genentech also performed some contract research for larger companies, which provided a source of revenues, and, like Syntex, set up joint ventures in projects outside its strategic focus. Also like Syntex, Genentech saw its capital structure and strategic plan as an integrated whole. Earlier products were licensed out, with a gradual integration of function—production, regulatory affairs, marketing—as new products moved down the pipeline. Each new product successively built up the base, both financially and technically.

What went wrong? There was a strong element of, at best, overoptimism, and, at worst, hubris. Genentech fulfilled many of its promises, only to be devastated by delays and deflations. These problems were most serious with a product that Genentech, and most of Wall Street, believed would be its version of the Pill, the blockbuster that would power it into the ranks of the pharmaceutical elite: tissue plasminogen activator, or t-PA, a protein that dissolved clots that cause heart attacks. The company first ran into delays at the FDA, which demanded more testing and more information. These delays kept Genentech's cost structure high—it had assembled an expensive marketing team in preparation for the launch—eroded its lead against competing products, and undermined its carefully constructed image. Problems with the FDA were not unusual, whether in biotechnology or the drug industry (Syntex also had a tough time getting new products through the FDA in the 1980s). But with its large expectations and an insurgent self-image, biotechnology brought a particularly nasty edge to its FDA relations. The FDA was overburdened and understaffed and was wrestling with new questions about dealing with biological products. Its style, as always, was fussily bureaucratic, in contrast to the aggressive entrepreneurialism of the companies. And while established drug companies had products already and could patiently wait upon the FDA, the biotechnology companies sat watching their money burn up as the agency pondered the situation. A few biotechnology companies, such as Genentech and Cetus, compounded the situation with a blend of arrogance and ignorance, making errors in the design and presentation of their clinical tests, then suggesting, particularly to the markets, that it was the FDA's fault.

Genentech lost almost a year because of delays at the FDA on t-PA.[25] The delays were particularly important because of the profound intricacy of the biology. Despite the hype, few biological products were truly unique, like nylon in 1939 and the silicon transistor in 1957. But patents did tend to be porous. Both the therapeutic targets (cancer cells, the immune system, the clotting system) and the therapies themselves (antibodies, proteins) were so complex that competitors were able to find new paths around, or through, patent defenses, stirring up messy, involved, expensive legal battles. Technical change outstripped scientific advance. Thousands of different antibodies could be developed, but their effect on, say, cancer cells, was problematic. Many companies developed ways of producing various forms of interferon—but to what end? The complex relationship of individual members of the interferon family within the immune system still required considerable unraveling. In the mid-1980s, researchers were still wondering whether the interferons were predominantly antiviral agents, anticancer agents, or both. Molecular biology was undoubtedly evolving forward. The field was large and global, and knowledge diffused rapidly, but the situation still resembled that of the drug industry in the early 1950s when the state of technical manipulation, based on organic chemistry, raced far ahead of biological understanding. The gap, in other words, had not closed.

Genentech's own expectations outstripped its resources. The company first exploited, then was victimized by, the potent feedback process among markets, the media, and corporations. The effect, as long as it lasts, is self-perpetuating, like conglomerate financing. Genentech, which won employee loyalty with generous stock options, accepted as fair the high stock price Wall Street rewarded it for most of the decade. Both the market and the company embraced the perceived potential of interferon—it would cure, said the promoters, the common cold and cancer. They then embraced the possibilities inherent in human growth hormone and, most spectacularly, in t-PA. Although at least a dozen other clot busters were under development by the latter 1980s, and skeptical murmurs were coming from the cardiology community, particularly in Europe, over t-PA, some Wall Street analysts still forecast that sales could go above $1 billion annually. Genentech, too, implicitly accepted those valuations by buying back the limited partnership that had funded development of the drug. In 1986 Genentech bought back the rights for some $360 million.

The reality of t-PA's worth was considerably less than valuations reflected in Genentech's stock price or the repurchase of the partnership shares. In early 1987 the FDA finally approved t-PA, under the trade name Activase. After a quick surge in U.S. sales, leading the company to declare it to be the fastest takeoff of any drug in history, growth abruptly slowed. A new wave of clinical results, particularly from Europe, suggested that the benefit of t-PA over streptokinase, a much less expensive product already on the market, was marginal (proof that the FDA had not been completely wrongheaded to ask for more evidence).[26] Meanwhile, a competitive clot buster called Eminase from the U.K. drug company Beecham underwent testing and began to appear in European markets. Much was made of t-PA's high cost—as much $400 a regimen—particularly as the U.S. health care system groaned under increasing costs. On Wall Street, analysts began to take a harder look at Genentech's product portfolio (particularly after the market crash of October 1987). Now that t-PA had lost its status as a blockbuster product, what did Genentech have to entertain the crowd? The answer was, after a decade of intense innovation, not much—or at least not much, not yet.

Genentech's ambitions, reflected in an impressive capital structure, now became a burden. The company's product portfolio was not strong enough to support its cost structure. The stock plummeted. As Wall Street's disenchantment deepened, Genentech's options in terms of raising new capital diminished and the allure of employee stock ownership waned. To generate new products, Genentech faced years of large R&D expenditures. And once its credibility was undermined, the company had no other choice than either to scale back or to seek out a better-financed partner that could support these technological dreams, just as Raytheon's Marshall felt himself forced to do in 1950 or Syntex had in the 1950s. Unlike the Raytheon board, Genentech actively shopped itself in America and Europe, eventually striking a complex deal with Hoffmann-La Roche. Genentech may well rebound with new, important, profitable products; Roche is certainly betting on that.[27] And indeed, biotechnology as an industry may eventually develop the momentum finally to dislodge the drug elite. But not yet. The first decade or so of biotechnology has demonstrated that, so far, biotechnology as a technology lacks the kind of innovational acceleration that allowed transistor companies to break the vacuum tube oligopoly. Biotechnology as an industry continues to be characterized by high capital costs, treacherous regulatory barriers, and difficult R&D hurdles.

Was biotechnology a failure? It was if you base your answer only on the early expectations created by the industry itself. As a loose and often inchoate industry, biotechnology has at least survived, with the public markets continuing to fund new entrants. Although once major names such as Genex, Cetus, Genentech, Centocor, Integrated Genetics, Genetics Systems, and Hybritech have disappeared or been absorbed by other companies, a few companies have scrambled past takeoff—if not all the way to the big pharmaceutical table. The best known, of course, is Amgen, which rode two major products to legitimate profitability: recombinant erythropoetin (trade name Epogen), a protein that stimulates red blood cell production, and granulated colony stimulating factor (trade name Neupogen), a natural protein that triggers white cell anti-infective activity. With sales of about $1.1 billion, Amgen still has a way to go to reach Syntex at $2 billion, but it is self-sustaining, fully integrated, and growing at a rate of 30 percent annually. Amgen, however, is an exception. Most biotechnology companies now accept their roles as intermediaries between the drug giants and the world of academic molecular biology, a situation very similar to the custom chip companies of the 1980s. Like many of those smaller chip companies, most biotech start-ups accept the realities of partial integration, usually retaining the highest value-added segment—the research itself (roughly analogous to design in the chip world)—while finding other companies that will test, manufacture, and market its new biological products. On the up side, a boutique strategy can operate on less capital, with principals giving very little equity away to investors; on the down side, profits will always be limited, and survival means that continual innovation must occur. Of course, even the boutique can dream of discovering the blockbuster, which may, as happened at Syntex and perhaps Amgen, fuel its rise into the ranks of the fully integrated pharmaceutical elite.

All that is not to say that the pace of fundamental change in molecular biology will not eventually accelerate so that smaller companies can create the economies to shatter the established hegemony of pharmaceutical scale and scope. But consider one cautionary note: the advantage of surprise is now gone, and molecular biology has now diffused throughout drug research, which is one of the reasons that R&D costs have risen for everyone and that consolidation among the drug elite continues. If a revolution does occur, it would be less a war between two technologies, as it often was in the 1980s—empirical screening versus molecular biology—and more a pure test of the competitiveness of two industrial structures—large versus small.

# 11

# Gravity's Rainbow

*There is no more pleasant fiction than that technical change is the product of the matchless ingenuity of a small man forced to employ his wits to better his neighbor.*

—John Kenneth Galbraith, *American Capitalism*

*It is the entrepreneurs who know the rules of the world and the laws of God.*
—George Gilder, *The Spirit of Enterprise*

## Bureaucrats and Entrepreneurs

The story of postwar American technology is certainly that of television sets and transistors, of microminiaturization and genetic engineering, of the rise of financial markets and venture capital, but it is also the tale of the historical decline of one complex of popular ideas and the rise of another. The postwar era can be divided into two broad periods characterized by different perspectives on how technology has fit into the American economy and ethos. As we've seen, the conventional wisdom that resulted from World War II was that only big business, and big government, could nurture modern technology and innovation. Big business had dominated commercial research and development since the late nineteenth century; and World War II, a preeminently technological war, had been won through the application of size and economic power marshalled by big government.

At some point, probably in the 1960s, the merits of size, of hierarchy, began to fade. Gradually, sentiment shifted toward smaller units, toward the entrepreneur, the inventor, the company as a corporatist ideal where all employees were creative and entrepreneurial. Size and reach, financial power and organization, became detriments rather than necessary prerequisites to innovation. Thus arose one of the ubiquitous clichés of the age. "Quick and agile," began a perfectly ordinary *Business Week* story about supercomputer maker Cray Research in 1990. "That's what big ponderous companies wish they could still be. So they diligently hack away at the bureaucracy, hoping to unleash the creative energy—since faded—that first brought success."[1]

By the 1980s, the ongoing struggle had crystallized into a Manichaean dualism that pitted the innovative entrepreneur against the presumed dead hand of the bureaucrat. It was a very subjective division. General Motors chairman Roger Smith could battle entrepreneur incarnate Ross Perot while advocating a sweeping program of robotization that he claimed would renew and transform the auto giant. The proliferation of computer and semiconductor companies in Silicon Valley in California and along Route 128 in Massachusetts, the rise of the biotechnology industry, was either a paradigm of a new, fruitful marriage between entrepreneurs and technology or a self-destructive orgy of capitalism. In time the distinction between entrepreneur and bureaucrat assumed moral overtones—sometimes explicitly. "Entrepreneurs, though many are not churchgoers, emerge from a culture shaped by religious values," wrote Gilder.

> The optimism and trust, the commitment and faith, the discipline and altruism that their lives evince and their works require all can flourish only in the midst of a moral order, with religious foundations. Secular culture has yet to produce a satisfactory rationale for a life of work, risk, and commitment oriented toward the needs of others—a life of thrift and trust leading to investments with uncertain returns.[2]

Where did these ideas come from? Forty years ago the notion of creativity and innovation bottled up by bureaucracy was a fugitive idea embraced mostly by a small group of maverick Austrians. In 1942 Ludwig von Mises, a leader of the exiled school of Austrian neoclassical economics who had settled in New York, wrote a thin volume called, simply, *Bureaucracy*, which sketched out the dangers of the bureaucratic state and argued that a corporate bureaucrat was one who did not seek to maximize profits.[3] Von Mises, needless to say, thought that such regressive behavior was increasingly the norm. Around the same time, Schumpeter, in *Capitalism, Socialism and Democracy*, described the transformative role played by the bureaucrat's doppelgänger, the capitalist entrepreneur. While both books, with their emphasis on the process of economic change, were reviewed in the professional journals, neither immediately spawned a broad intellectual movement.

Economists, enthralled by Keynesian macroeconomics and soon to be captured by the powerful possibilities of the computer, were focused elsewhere.

With a few exceptions, the emphasis, both in economics and among the burgeoning population of management theorists, remained on the merits of size, on organizational adjustment, and, more technically, on macroeconomic equilibrium analysis. In 1954 Drucker's magisterial *Practice of Management,* with its articulation of management by objectives, represented a sort of ideal: large organizations that nonetheless transcend the social and economic divisions of the past and provide a satisfying, fulfilling environment for workers.[4] In a typical study of research management by two Harvard Business School professors in 1963, the two words *bureaucracy* and *entrepreneur* never appeared. Faced with the issue of how to encourage corporate innovation, the authors nervously touched on issues of psychological adjustment, contrasting "authoritarian" personalities with managers with "understanding" and "intuition."

> We were not intentionally seeking examples of authoritarian administration but we did encounter a number of instances of it in laboratories maintained by some of the largest corporations. On the other hand, we were impressed by the far greater number of research managers who brought to their jobs what seemed to us at the time a high degree of understanding of human motivation—particularly a good understanding of the needs of scientists trying to be productive in what many of them felt was a somewhat alien environment. This understanding, however, as in so many areas of management, seemed to be intuitive rather than explicit.[5]

In economics, the notion that size and innovation were inextricably linked proved to be a sturdy one. In 1952 Galbraith first declared the obvious truth of this notion in his *American Capitalism;* in 1967 he was still elaborating when he sketched out "the imperatives of technology" in *The New Industrial State.* Galbraith used as his prime example the automobile industry, specifically the development of the Ford Mustang in the early 1960s. In reality, the complexity and technological imperatives of designing and building a new car *did* require a large and complex organization. In great detail, Galbraith describes how the car companies become increasingly inflexible as the technology grows more sophisticated. Development time expands, costs rise, flexibility declines as the task is narrowly subdivided, specialized manpower has to be hired, organizational experts are consulted, and planning takes on supreme importance. One can almost hear the word *bureaucracy* murmured here, though Galbraith hews to his own, less derogatory nomenclature, with its allusions to Veblen, when he refers to a technostructure, a small ruling body of skilled experts from big government and big business.

> The more sophisticated the technology, the greater, in general, will be all the foregoing requirements. This will be true of simple products as they come to be produced by more refined processes or as they develop imaginative containers or unopenable packaging. With very intricate technology, such as that associated with modern weapons and weaponry, there will be a quantum change in these requirements. This will be especially so, as under modern conditions, cost and time are not decisive considerations.[6]

Galbraith concluded that only the state can provide the funding and planning necessary to push long-range sophisticated technological programs; it was a mentality that had produced the Manhattan Project and the space program. The technostructure dominated the modern American economy—thus, Galbraith urged a more enlightened attitude toward government planning and more intelligent regulation of corporate activities, foreshadowing the cries for industrial policy in the 1980s.

Galbraith's view of the dominance of large corporations—of large groups generally—was widely accepted, though many of his other ideas remained controversial; it was, to borrow the title of a Galbraithian collection of essays from a few years earlier, "the liberal hour," and the belief in size was a keystone of cold war liberal orthodoxy. There was, of course, a measure of truth to it. Among many of the largest firms, particularly in the defense industry, such a technostructure existed then and exists today; and embedded in such a structure, the market factors of "cost and time" *are* less important. But what is most striking about his comments when one reads them today is what he ignored. Galbraith discounted Schumpeter's analysis of the dynamic character of technological change. Galbraith's New Industrial State had a static quality, as if large corporations had achieved the means to repel all attacks, for all time. He confused the intricacy, scale, and expense of carmaking technology with its maturity. The automobile was a nineteenth-century invention, an advance related to that Ur technology of the industrial revolution, the steam engine. The automobile, by the 1960s, had, for a variety of reasons, long matured. Improvements came incrementally; many—if not most—were cosmetic in nature, and all were expensive. Carmaking had more to do with engineering than with science, more to do with manufacturing and marketing than with basic research (a situation that the Japanese exploited). The Mustang involved no fundamental advance beyond the technological past; it represented a trend toward simpler, sportier, smaller cars for a demographically younger market. The Mustang was simply a continuation of GM chairman Alfred Sloan's notion that different classes of people want different kinds of cars that periodically change their looks, like models at a fashion show. The structure of the industry reflected that maturity. The Big Four carmakers at the time (Ford, GM, Chrysler, and American) existed in a stable oligopoly. Except for minor nibbling by Volkswagen, foreign competition in the 1960s was as laughable as a Japanese compact car.

Galbraith thus captured a large, declining segment of the economy but missed a smaller, growing part. While he was pondering Ford, other large companies were demonstrating an entirely different economic paradigm—Fairchild Semiconductor and Texas Instruments with their integrated circuits, IBM with its room-sized mainframes, and Digital Equipment with its minicomputers. These companies mined the most advanced of basic science, and innovation and improvement came in great leaps. The industries were fluid, volatile; prices descended in long, sweeping curves. Cost and time meant everything. In these organizations the technological imperatives had not yet set; these new companies could make and sell products and grow so quickly that they beat back the established giants that to Galbraith controlled the marketplace like lords of the manor. They embodied conflict: new versus old, small versus large, entrepreneurial versus bureaucratic, creativity versus authority. Even IBM, which in the 1960s dominated its market and represented the apotheosis of Drucker's corporatism, with its seeming blend of entrepreneur and manager and its policy of lifelong employment, found itself furiously competing, and losing, as the technology spun on. In industries undergoing rapid technological change, technological entrepreneurs quickly found themselves masters of large organizations. Such dramatic successes were increasingly celebrated as an internal state of entrepreneurial grace, a spirit of renewal and revolution, and they were cheered by the stock market and the media.

By the 1980s American business had embraced the dream (if not quite the reality) of transformation and renewal, though even its most fervent proponents were quick to distinguish their commercial or technological insurrections from the sordid, political variety that erupted in small, humid, foreign countries. Drucker provided a variation on that archetypical American dream of constant change without pain in his bestselling 1985 book, *Innovation and Entrepreneurship*. Wrote

Drucker: "They [entrepreneurs] achieve what Jefferson hoped to achieve through revolution in every generation, and they do so without bloodshed, civil war, or concentration camps, without economic catastrophe, but with purpose, direction and under control."[7] Thus, entrepreneurship had become not only good but deeply American. And Drucker's entrepreneurial revolution occurred more frequently and created less havoc than that of Schumpeter. By then the balance had decisively shifted from the merits of the large business organization to that of the entrepreneur.

The stark division of experience—bureaucrat versus entrepreneur—thus arose from the widespread belief that the American economy was in the process of being reordered through the instrumentality of technical change. This view found its most articulate apostle in Gilder. A former Nelson Rockefeller and Ronald Reagan speech writer and free-lance thinker, Gilder proved to be nearly as popular, and as controversial, as Galbraith, while taking a diametrically opposed view. What mattered was not big business but small business; entrepreneurs, not bureaucrats; technology, not management. If entrepreneurs were actors in a technological revolution, then bureaucrats were keepers of the status quo, a narrow-minded mandarinate of office holders and salarymen. Entrepreneurs were creative; bureaucrats were dull, stultifying. In a world that was changing so rapidly, managers had to change just as quickly; indeed, in an amoral world, it was about as close American business came to a moral code. And it was, at least superficially, a radical moral code at that.

Galbraith and Gilder both produced clear, bold, prescriptive answers to the large question of where technology fit into the postwar American economy. Between them runs the spectrum of popular sentiments about technology and business. One was nominally liberal, a planner; the other was nominally conservative, a man of the market. One emphasized size and organization; the other preached the gospel of creativity and freedom. One argued the determinism of the "technological imperative"; the other told of the power of an inner light, of mind, of the "silicon imperative." Both found themselves defending ironic positions. Galbraith, the liberal, emphasized stasis; Gilder, the Reagan conservative, emphasized personal, commercial, and national transformation. Both, significantly, have had more influence politically than academically. And both, finally, shared a belief that technology is a substance with monolithic tendencies and that the means to unleash it are similarly simple, straightforward, and obvious.

## The Productivity Conundrum

In *The Post-Capitalist Society,* his twenty-sixth and most recent book, Drucker, now seventy-four, still argues that the key to this new age is the mastery of organizational management he takes credit for pioneering four decades ago.[8] Despite confidence in the wonders of management, Drucker offers up a formidable laundry list of fundamental questions that need to be answered if we are to prosper in the new century and the new age where entrepreneurs and knowledge—not capital, labor, and materials—dominate.

One of Drucker's questions concerns the conundrum of lagging productivity—a raw wound that undermines claims for a new technological era full of creative, happy entrepreneurs. Simon Kuznets was right in the early 1960s to worry about productivity; he was just a bit early. In 1973—several years after the invention of the memory chip and microprocessor supposedly decentralized intelligence, and the very year of the first genetic recombination—productivity growth abruptly slowed in the developed world, particularly in America. No one knows why, though offering explanations has become a minor academic industry. Commentators initially blamed the slowdown on the first oil shock, which sent the economy plunging into recession in 1973, or the collapse of the Bretton Woods arrangement in 1971, which freed foreign currencies to float against the dollar. But even when oil prices again fell (and international trade grew, rather than declined, despite currency volatilities, in part because of sophisticated hedging instruments developed in American financial markets), labor productivity growth failed to rebound to the 3 percent or so average between 1950 and 1970. Instead, it limped along at 1 percent or so—a dramatic difference. If the economy had grown at the 3 percent rate since the early 1970s, productivity—and, roughly, incomes—would now be almost 50 percent higher than it is. Other measures of productivity show a similar pattern.

The productivity slowdown soon generated political comment. Conservatives at first blamed inflation, high taxes, onerous regulations, and Jimmy Carter. Alas, those explanations lost their force as inflation, federal taxes, and regulation fell and productivity failed to respond. Even when the Reagan administration cut taxes and loosened regulation, productivity barely budged. Liberals then began to argue that productivity was being smothered by the upsurge in takeover activity, the deepening federal deficit, and the deadening weight of the cold war defense establishment. By the mid-1980s, American education was absorbing some of the blame, as test scores dipped, particularly in science and mathematics. Observers pointed to a decline in patents taken out by American corporations, and a mounting percentage of foreigners taking engineering degrees in U.S. universities. American business schools were blamed for teaching financial manipulation over manufacturing, corporate gamesmanship over entrepreneurship. Such arguments easily slid into a cultural critique: Americans were simply too fat and lazy, too dumb, too narcissistic and hedonistic to compete in the global economy. Managers blamed workers. Workers blamed management. Everyone blamed the bureaucrats. So much for the corporatist dream.

Economists explored more technical explanations. Edward Denison, a Brookings Institute economist who picked up the study of productivity where Kuznets left off, has worked over the years to account for productivity through various economic inputs: hours worked, rising education, rising population.[9] In the late 1970s, Denison focused on a broad area left over after adding up all quantifiable inputs. He called it "advances in knowledge"—essentially, the progress made because of technological advances. By the late 1970s, conventional wisdom, particularly in the scientific world, blamed the general slowdown on a real decline in R&D spending through the 1970s and 1980s, after three decades of massive increases. Denison attacked the question. Indeed, R&D spending as a percentage of gross national product had grown from 0.95 percent of GNP in 1955 to peak at 2.97 percent in 1964. By the late 1970s it fallen back to 2.77 percent, mainly because of cutbacks in defense and the space program, two areas, Denison admitted, "whose connections with productivity [are] slight." Expenditures from other sources—mostly industry spending on R&D—continued to mount through the 1960s. The problem, as Denison says, is that there is no indication that an economy twice as big will require R&D twice its size. His conclusion: slower R&D spending growth had *little* effect on the productivity slump.

Denison also reeled off other explanations for the decline, each of which proved even more difficult to prove. First, there's the cultural question—what he called the decline in Yankee ingenuity. The argument, however, tends to be circular: the evidence of a decline in ingenuity springs from the decline in productivity. Second, he examined the claim that slowing productivity is the result of a decline in opportunity for new advances. The technological economy had exhausted the easy opportunities unleashed by World War II and the subsequent organization of science and technology: the technological economy, in other words, had matured. This theory, Denison noted, stems from another of Schumpeter's favorite notions, borrowed from the Russian economist Nicolai Kondratieff, of long-wave cycles of innovation (other observers point to other long-term cycles, such as credit cycles). Evidence against the wave argument, Denison notes, is the suddenness of the decline. Why did productivity plunge so suddenly in 1973 rather than gradually taper off?

By the 1990s, another explanation loomed: the productivity slowdown involved a structural metamorphosis in the economy. During the postwar period, the economy has steadily shifted from one based on manufacturing to one dominated by services, just as the dramatic shift from agriculture to manufacturing had occurred earlier in the century. Indeed, that very decline was a major reason for what sociologist Daniel Bell called postindustrialism. The dominance of services accelerated in the 1970s and 1980s: services accounted for 55 percent of the jobs in the private economy in 1970, fully 75 percent two decades later.[10] Manufacturing industries such as machine tools, consumer electronics, automobiles, and heavy equipment ran into international competition first, suffering particularly grievously in the early 1980s before cost cutting and new technology revived its productivity and its competitiveness. At the time, however, some observers were prepared to sacrifice manufacturing on the altar of postindustrialism and allow it to wither away or be transferred out to low-cost, Third World regions like Southeast Asia or Mexico. Manufacturing was dirty, rigid, and full of blue-collar unionized workers, a remnant of the past; services was clean, flexible, a refuge of knowledge workers or symbolic analysts, a sign of the future.

Alas, the service sector not only slammed into renewed competition, but it proved slower than manufacturing to boost its efficiency and productivity. Here the mystery deepens. After all, few sectors more quickly sought out automation than services; IBM itself began as a company making tabulating machines for office workers. But despite massive spending on computers and telecommunications, service operations continued to hire more bodies; the ratio between back-office employees (that is, the support staff, from clerks to telephone operators) and front-office employees barely budged in the 1980s, despite extensive computerization and a dramatic surge in employment (service companies accounted for most of the twenty million jobs added to the economy in this period).[11] Moreover, the technology often failed to spark the significant improvement in efficiency, as Stephen Roach, the chief economist at Morgan Stanley & Co. and an acute observer of service-sector productivity, wrote in 1991:

> Productivity benefits from information technology in open-ended office applications have been especially elusive. For example the seemingly attractive and very expensive concept of the fully net-worked office rings increasingly hollow. Technology connects

machines, but so far it has done little to instill productive synergy among people. An interconnected office environment may facilitate the flow of electronics messages, but the creative high-value-added applications are still lacking. And yet it is precisely those sorts of innovative, idea-driven breakthroughs that lie at the heart of America's long history of productivity enhancement.[12]

Similar productivity problems continue to persist in huge service industries such as education, government, and health care.

Unlike manufacturing, however, service firms were still insulated from foreign competition through the 1970s—because of regulation (airlines, financial services) or because the players were traditionally local (health care, education). But once deregulation ensued and international competition began, the weaknesses were exposed. The infrastructure of automation installed over the past few decades ironically made the ability of the service sector to adjust and compete so much weaker—you can't fire a computer. As competition mounted, service operations began to lay off huge numbers of employees. They began the hard work toward greater productivity and, significantly, greater scale. In fact, contrary to the gospel of decentralization, service companies—the dominant organizations of postindustrialism—also compete on economies of scale just like the giants of industrialism. And service firms, like their industrial counterparts in the past, are also the closest thing we have to classical bureaucracies.

Roach finds an optimistic scenario in this gloomy diagnosis. America is the most advanced service economy in the world, and it is suffering through changes that other countries would have to undertake. Moreover, in the face of very deep cost cutting in the early 1990s, and the application of new technologies, particularly to so-called knowledge workers, Roach finds deliverance. Service companies are moving up the line from automating the back office to developing ways to "leverage" creative front offices. Outsourcing has reduced fixed costs; new flexible personal-computer-based information technologies have come on-line, and Roach predicts that with hard times forcing companies to reduce payrolls that the first service productivity recovery has begun.

Pinpointing the service sector still does not fully satisfy the desire to discover more fundamental reasons for why American companies haven't been more competitive, why they have not been managed more effectively. In the soul searching of the early 1990s, a number of explanations were offered. First, there was the claim that companies no longer invest in new research, technology, and new product development because capital is so expensive.[13] The argument was that the cost of American capital had soared well beyond that of the Japanese, deterring American managers from investing in long-term, high-risk projects, from semiconductor memories to television sets. Critics pointed to Americans' low national savings rate, arguing that in the 1980s, Americans went on a consuming binge while the Japanese continued to lead the world in personal savings. Unfortunately, productivity was flagging in Japan as well. Moreover, venture capital and initial public offerings in the 1980s hit all-time highs—though one school of thought argues that the very style of U.S. venture investing itself encourages short-term, speculative tendencies. While a number of studies seemed to confirm what corporate executives anecdotally report, more recent studies have indicated that the cost of capital issue is far more complex than originally thought and that managers' perceptions may be distorted. Not only are national comparisons difficult to make (and they miss subtle differences that can raise the real cost of capital above nominal market rates), but the explanation fails to account for the fact that American corporations may have had the cost advantage in the 1960s, when capital in Japan was relatively scarce, and the 1990s, when U.S. interest rates fell and the Japanese bubble economy collapsed.

A broader variation on the cost of capital issue is the short-termism charge. Highly speculative American capital markets have forced corporate managers to invest for the short term, threatening those corporations that fail to perform with takeover and dismemberment. Once again, the empirical evidence is sketchy. Studies instead showed that U.S. capital markets tend to reward companies that announce capital expenditures and increases in R&D. One explanation is that U.S. markets are generous with high-return, emerging companies but, as in the television-set industry, are quick to abandon a broad middle tier of companies that may be aging but still have considerable life left in them as long as reinvestment occurs (these firms also provide a major chunk of employment). By the early 1990s a series of task forces and reports began to urge legislation that would encourage less speculation, less market turnover, and longer, more permanent ties between still-growing U.S. institutional investors and U.S. corporations. The argument, ironically, goes back to Berle and Means—but with a twist. Like Berle and Means, a number of commentators argued that the continuing separation of management and ownership has led to corporate drift, and to bureaucracy. Their solution: seek the same kind of close, financial ties between investors and corporations that supposedly exist in Japan, with its *keiretsu* groups, and Germany, with its *grossbanken,* or large banks, and encourage a more egalitarian participation of workers, managers, engineers, and researchers through quality circles and flexible manufacturing techniques, often again derived from Japanese practices. Ironically, these same arguments tend to downplay the other part of Berle and Means's depression-era message—that large economic units were stifling capitalist energies—and ignore the contemporary gospel of the small, entrepreneurial unit as well.

## Beyond Economics

Increasing concern over productivity produced two political solutions in the 1980s: supply-side economics and industrial policy. As MIT economist Paul Krugman has noted, both prescriptions lost favor in the late 1980s, supply side because of disenchantment with the Reagan years and industrial policy because of the pervasive and continuing distrust of government. Krugman, one of the bright young stars of economics, lacks the optimism of his Keynesian forebears about solving macroeconomic problems—particularly productivity.

> There are various things the government can do that might accelerate productivity growth without political risk, from encouraging higher education standards to supporting a few industrial consortia. These things will be tried and perhaps they will even work. But the basic political consensus at present is that a low rate of productivity growth is something America can live with. . . . Productivity growth is the single most important factor affecting our economic well-being. But it is not a policy issue, because we are not going to do anything about it.[14]

Krugman makes the point that many advocates of schemes to revive productivity are either economic heretics or noneconomists: from economist Arthur Laffer and journalist Jude Wanniski on the supply side, to lawyer and political economist Robert Reich and consultant Ira Magaziner on industrial policy. Krugman's comments may reflect some envy from the technical economists who found themselves in political exile during the Reagan and Bush years, but they also reflect, in his own phrase, "the diminished expectations" toward real change by the public and by economists themselves (the public seems to share the diminished view of economists as well). That trend of leaning on economic outsiders continued into the Clinton era. Clinton named old Oxford classmate Reich to be his secretary of labor; Laura Tyson, a slightly heretical trade economist from Berkeley, to head the Council of Economic Advisers; and Robert Rubin, a lawyer and former head of Goldman Sachs & Co., to serve as his economic security council chief. Krugman, piqued, loudly complained.

Confronted by mediocre productivity figures and the limitations imposed by technical economists, apostles of technological and economic progress have been forced to take more extreme positions. The cry of full employment and growth that was heard in 1946 became the gospel of transformation and revolution in 1992. Schumpeter, not Keynes, is now the touchstone. Entrepreneurs must be encouraged or America will slide into a no-growth, make-work, socialist economy—like Britain before Margaret Thatcher. A new age is dawning that is qualitatively different from the past, with different rules and a new kind of man. The past is, for the most part, irrelevant to a fast-moving future.

Talk of transformation, of revolution, produces a heterodox faith, one that is often antagonistic to technical economists, with their interest in a quantifiable past. No one has articulated that faith with more fervor than Gilder. In *The Spirit of Enterprise* (1984), Gilder attacked nay-saying technical economists as a sort of organized army of conventional wisdom. Gilder charged economists with focusing exclusively on a world made up of "colliding multinational corporations, national industrial policies and macroeconomic tides that overwhelm the simple energies and enthusiasms of the individual entrepreneur"[15]—in other words, blaming the economic physician for the disease. Gilder, in a bout of neo-Platonism, then went on to argue that the world of macroeconomic forces and giant institutions is less "real" than that of the nascent, emerging world of entrepreneurs driven by an emotional fuel of questing, wanting, willing.

Gilder was, of course, a supply sider. He argued that the capital gains cuts of 1978 and 1981 had released so much pent-up innovation that, combined with a rising Reagan stock market, it would liberate the economy from Carter-era doldrums and revive productivity. "While the leading experts blindly sang a statistical dirge of productivity and stagnation—and leading economists predicted depression—entrepreneurs, spurred on by the drop in capital gains taxes, launched a broad economic revival," wrote Gilder in his 1989 book, *Microcosm.*

It is a country of entrepreneurs who began a profound and far-reaching economic revival in the late 1970s that addressed and resolved many of the problems that the critics cite. It is a country of small firms and new inventions and radical breakthroughs by daring men who are leading the world economy into a new age of growth and prosperity. Indeed the economy of entrepreneurs can make a plausible claim that they, rather than the world of econometrics, are the real economy.[16]

Although the stock market advanced, productivity failed to budge, though Gilder slips off the hook by blaming the Bush administration for derailing the revival by raising taxes. If you turn the temperature down a bit, Gilder, like Galbraith thirty years earlier, has some valid points to make. The economic performance might have been worse if the broad expansion in venture capital, the public markets, and junk bonds hadn't produced the capital for many new companies. And the reaction against takeovers and "paper entrepreneurialism" in the 1980s may have been overstated a bit; the record is mixed and takeovers arguably produced less lasting damage than the strong dollar and high interest rates of the early Reagan years. On the other hand, the growth in defense spending also inflated economic indicators like the stock market and employment without fundamentally affecting underlying productivity, and the tax cuts helped to expand the federal deficit dramatically. Meanwhile, a considerable amount of capital and initiative went into takeovers and leveraged buyouts and the huge savings and loan and real estate disasters.

By the late 1980s, Gilder had zeroed in on what he viewed as the key to the coming entrepreneurial transformation: the dramatic explosion in the power of microelectronics. Gilder builds his argument around rough precepts about microelectronics formulated by Carver Mead, the Caltech semiconductor guru. Like the Heisenberg uncertainty principle, Mead's rules for operating at the quantum level are counterintuitive: as you drop deeper into the microcosm—as chip components get smaller and smaller—chips get faster, cooler, cheaper, more spacious. Gilder argues that capital, materials, and labor diminish in significance (odd, since just a few years earlier he argued that the revolution was occurring because of a tax cut that would free up capital) and the industrial hegemony fragments. Although most of *Microcosm* is a fascinating, if quirky, history of microelectronics (as we saw, the structural consequences of these rules are far more ambiguous than he suggested), Gilder ends with a blast of metaphysical trumpets, sounding like Buckminster Fuller one moment, Nietzsche the next. The microcosm will slay the dark forces of scientific materialism and determinism that have ruled since Newton and create a new synthesis. "The quantum vision finds at the very foundations of the material world a cross of light. Combining a particle and a wave, it joins the definite to the indefinite, a point of mass to an eternal radiance. In this light we can comprehend the paradox of the brain and the mind, the temporal and the divine, flesh and word, freedom and fatality." God, in other words, is an *American* capitalist entrepreneur.

Gilder is something of an acquired taste, but he is representative of a current predisposition to see science and technology as the key to history. In many ways, he resembles a figure from the nineteenth century: Herbert Spencer, the civil engineer turned journalist turned philosopher who seized upon Darwin's evolutionary theories to build an edifice we now know as Social Darwinism. Like Spencer, Gilder extrapolates from fundamental scientific advances, attempting to recast men in the light of recent scientific "truth." Both believe that humankind attains its greatest authenticity in the hurly burly of the market; and both work to reconcile men and women to a harsh, competitive existence of struggle and change. Although Gilder decries the effects of materialist determinism, his scheme, like Spencer's, has its deterministic tendencies. His emphasis on the power of the quantum idea, his notion of a "real," if not yet quite established, economy, his argument that we have entered a new age, smacks of those old determinists Hegel and Marx (Gilder might reject scientific materialism, but he still embraces one of its greatest triumphs, quantum physics, as a catalyst of the new age, a view that resembles Marx's admiring view of creative if destructive capitalism as a necessary precursor to socialism). In Gilder's world, scientific and technological progress triggers a larger transformation of humankind, of morals, of business, of economics, of history; and America, with its individualism, its Protestant and revolutionary roots, its Emersonian tendencies, is far more of an archetype for the microcosmic era than the Japanese, with their hierarchical and communal tendencies. Gilder condemns industrial policy and rejects foreign models in both finance and management. Free markets, political and financial, represent the highest virtues of American individualism—a style, he suggests, that is spreading around the world as the struggle to innovate and develop mounts. The very nature of the information, or quantum, revolution is peculiarly American. From there, it is a short step to arguing the simplistic effect of information technology on grand political events—notably, that faxes, satellite television transmission, and computers (from Star Wars to PCs) brought down the Soviet empire and ended the cold war.

Gilder believes in, and predicts, the advent of a new stage, a new age, spawned by science and technology, dominated by America. With the cold war over and the millennium approaching, with continuing technological change, the urge to end one chapter in human endeavor and begin another waxes strong. Gilder has his microcosm; Alvin Toffler, his Third Wave; Drucker, his postcapitalist and knowledge society; Daniel Bell, his postindustrial society; and Francis Fukuyama, his Hegelian end of history, with its triumphant rise of liberal democracies and its resulting, very Schumpeterian, malaise. The list goes on and on. In his book, *The Control Revolution,* James Beniger lists some seventy-five declarations of ongoing social transformations between 1950 and 1984.[17] They all differ in scope, sophistication, and politics. What they share is a sense that the world has qualitatively changed, mostly through the instrumentality of technology, and that we have entered a transitional phase leading to a different (some say terrific, other say awful) stage of progress.

## The Age of Clinton

All that may be true; we will have to wait. But what strikes one most dramatically after exploring the recent technological past is not the differences and discontinuities but rather the extensive uniformities that exist between industries and technologies over time. Certainly, according to the very narrow definition of technological revolution that I offered at the beginning of this book—a technology and its corporate agents that overthrow a previous technological regime—we have seen at least one such revolution: the dramatic replacement of the vacuum tube by the transistor. We have also seen the disruptive effect of innovation on mighty corporate structures: Philco, Admiral, and RCA no longer exist as independent companies; DEC and IBM, so recently so dominant, flounder; and, almost daily, the corporate nameplates of the past disappear to the sideline chants of creative destruction. But again, while technology plays a role in this (and the notion of technology as the great determining factor of history is another great cliché of the age), it is not the only factor. The growth and decentralization of financial markets and the globalization of manufacturing and trade are just as responsible for turmoil among large American corporations as pure, technological change. The urge to reduce everything to technology—the chip or genetic manipulation as transformative agents—is very strong.

Revolutions, however, have a way of submerging back into history. The saga of microelectronics, while dramatic and full of upheaval, increasingly resembles that of other great industries. Today there are even flashes of maturity in microelectronics, a return perhaps to the larger, dominant corporation, to oligopoly and market control, albeit undoubtedly with a less firm hand on product markets. Is Intel the new GE? Will the new AT&T revive the natural monopoly of the old AT&T? At least the possibility exists. All that one can really say is that retaining a grip on markets for now is far more difficult than in the past.

Although the triumph of smaller, entrepreneurial companies seems complete, the strengths of the large corporation stubbornly survive. Although technological change has limited the old industrial economies of scale, they reappear, often in new guises, such as the learning curve of

microelectronics. Moving over the landscape of the technological economy, it is difficult to discern a brand new age in which economies of
scale are permanently banished, in which information alone rules (just as
it is difficult to find an earlier age of capitalism—certainly not the nineteenth century—when knowledge and information were not essential for
innovation).[18] Instead, these polarities—centrifugal and centripetal,
large and small, industrial and postindustrial, capital and knowledge—
seem to oscillate continually as the pace of technological change varies,
as the nature of regulation and the configuration of product markets
change. In some highly technological industries, scale and scope appear
to be gaining, not losing, in importance. In the 1980s, the drug industry
managed to co-opt biotechnology, adapt the new methodology of molecular biology, and refit itself for global competition. All those factors
drove pharmaceutical operations to seek greater critical mass through
consolidation; a similar tendency exists in consumer electronics, particularly in Japan and Europe. To view biotechnology as the future of pharmaceuticals is simplistic, just as it is unwise to predict the demise of the
dominant software producer (Microsoft) or semiconductor manufacturer (Intel) or, more to the point, all dominant companies.

How does this fit with the plans that the Clinton administration is
making to dress America up for the twenty-first century? The administration faces some larger problems that limit its ambitions. The demands
of deficit reduction, a still-sputtering economy, and a much-reduced
defense establishment put pressure on federal funding; major battles
have already broken out over Big Science projects, such as the supercollider and the space station, that the administration has tried to cut back.
Competition and antitrust have forced reductions at two of the greatest,
privately financed, "national laboratories": Bell Labs, now focused more
tightly on commercial technologies, and IBM Research.[19] The United
States will certainly not return to the funding levels for basic research
that existed in the late 1960s, before the great inflation, until the deficit
problem is solved. Moreover, post–cold war attempts to shift defense
R&D to civilian purposes (such as removing the "Defense" from the
Defense Advanced Research Projects Agency or pressing for "strategic
research") sound good in theory but are more difficult in real life. Agencies like ARPA operate by choosing technologies and industries, which
inevitably raises all the questions about an industrial policy in a civilian
economy that is a lot more complex in its structures, goals, and use of
technology than is the military. The government's record in deliberately
picking technologies in the past has been lousy.[20]

The administration also seems to lack a clear sense of its own mind on the question of corporate structure and innovation; it seems to believe that one can simultaneously please two groups of companies, the large *and* small, the old and the new. Clinton talks stoutly of defending American interests abroad (which usually means helping large American companies, from Boeing to Intel), but at the same time he preaches aid to small businesses. Alas, the very act of succeeding with one group may damage the other; the civilian economy is fraught with zero-sum games. Encouraging competition can only be a boon, as long as that market competition, with winners and losers, is viewed as a healthy phenomenon, not one to be locked in a room with a guard on the door.[21] Promoting consortia, encouraging alliances, and suspending antitrust only give more advantage to already dominant companies. Negotiating a new semiconductor agreement with the Japanese could well hurt smaller computer makers by forcing them to pay higher prices for chips.[22] And, as we have seen, domestic competition and a multitude of self-imposed wounds, not foreign devils, have brought down large firms such as IBM and DEC.[23] Social imperatives also conflict with technological goals. The administration has talked of regulating drug prices—an old issue that just won't go away—and reducing pharmaceutical profits. But taking pricing freedom from the large companies will hurt the small biotechnology firms even more, particularly if the administration takes no active steps to accelerate the FDA approval process and imposes new layers of regulation.

The administration's actions also suggest a lack of sophisticated thought about the interaction of various funding sources. Federal funding for basic biological research has done more than anything else to close the gap between medicine and pharmacology. Despite the hype, that gap is still quite wide; and there remains an essential role for truly basic research, unconnected with commercial considerations. There is also, needless to say, great demand for applied work, but that is being filled by pharmaceutical R&D budgets and by the biotechnology companies, funded generously, if intermittently, by Wall Street. Shifting the National Institutes of Health to more focused, more "applied" research—implicitly choosing targets, often to achieve social, not scientific, goals—threatens to further demoralize an institution that, in biology, once rivaled that of Bell Labs in electronics and further drive academic research into the arms of commercial interests. Targeted biology is also fraught with uncertainties. When Clinton talks about a "Manhattan Project" for AIDS he is of course making a political statement, but he is also implicitly comparing apples (quantum physics) and oranges (molecular biology); the quantum model was complete, the model of the immune system is not. A Manhattan Project for AIDS may advance research into the deadly disease, but, like the war on cancer, it will also dramatically raise expectations that may be impossible to meet.[24]

Harkening back to the Manhattan Project is symptomatic of the administration's general bias for technocracy over markets. The administration believes that it can process information and choose an R&D agenda far better than the interplay of many self-interested "investors"— whether narrowly defined as buyers of stocks or broadly defined as the foot soldiers of self-regulated academic science—that makes up a marketplace. Despite the rhetoric, this issue is not at all clear. Markets are clearly imperfect, and prone to fashion and fad. But they are protean, flexible, and fast. A permanent technocracy can, like Vannevar Bush's OSRD, act wisely and effectively; but it can also be inflexible, shortsighted, slow, and dumb. Belief in a technocracy suggests that we have mastered the inner mechanisms of innovation and economic growth; it reflects either Clinton's optimism about human capacities or hubris about his own undoubted intelligence.

It is natural to react to the immediate past. The Clinton administration reveals, in a variety of ways, the American obsession with Japan in the 1980s—and the complementary despair over American prospects. There are many good reasons to look at Japan with admiration: its manufacturing techniques, its drive for exports, its discipline. But there has also been a tendency to exaggerate Japan's strengths—and to ignore her weaknesses. In many ways, the Clinton administration remains fascinated with the image of Japan's Ministry of International Trade and Industry (MITI), that hive of master bureaucrats shrewdly picking and choosing technologies and systematically undermining American efforts. But few pause to remember certain realities. First, MITI's power has waned as Japanese corporations have prospered, loosening some of the tight *keiretsu* relationships of the past. Second, MITI failed as often as it succeeded; and some Japanese triumphs came from ignoring the bureaucrats. Perhaps most important, catching up is a far simpler process than forging ahead—a matter of engineering (or reverse engineering) over science, manufacturing over invention. To model American technology policy on Japanese policy is to confuse the two, to cut the power on the very engine that keeps the United States ahead—that is, the basic science pursued in national labs and in academia.[25]

There are a number of Clinton initiatives that may be more helpful over the long term, such as funds for an information infrastructure, the famous data super highway. But there is one broad set of goals that might do more to aid the American technological establishment, particularly its commercial end, than any series of glittery projects. More than anything else, companies, large and small, crave stability, not only in terms of the macroeconomy (interest rates, the markets, currencies, inflation) but in terms of regulation and, particularly, tax policy. Continual shifts in policy extract a high, if undefinable, cost. One of Clinton's most vocal promises during the 1992 campaign was for an investment-oriented administration dedicated to long-term change. Long-term investment, as Keynes recognized, can take place only when there is some confidence in long-term prospects and a long-term environment. And no single institution plays more of a role in determining that environment than the government.

## An Ecology of Technology

History, writes John Lukacs, does not repeat itself, it forms new combinations. That is certainly true of the technological economy. The system that has developed in the United States is one of great complexity, constant change, and feedback. Quantification is difficult (though, by now, much effort has been put into quantifying technological inputs and outputs), and one is left with description and portraiture. The technological economy consists, like the ecology of the woodland pond, of cycles within cycles—in science, in technology, in the macroeconomy—and of a myriad of external determinants that is as much the province of the historian as the economist: political change at home and abroad; trade law; tax reform; the indeterminable effect of personality and genius; the role of fad, fashion, and memory. Such a system may well move in long rhythms, as Kondratieff said, though one does wonder why over long periods of time these intervals are so rigidly spaced. If there is more capital for science and technology, if we are somewhat more self-conscious, more rational, about stimulating innovation, why should we be locked into the same rhythm as in the late nineteenth century?

Cyclicality implies limitations, reversals, inescapable cycles of birth, growth, and death. It suggests that progress moves backward and forward, that innovation and change accelerate and decelerate like a car in traffic. Technology in the end may be a human artifact, like markets, but it is also part of nature. The gap between science and technology closes; innovation accelerates and then slows as maturity sets in. New technologies blossom quickly, bloom, and succumb to inevitable decay, tracing a sine curve across time. Then, as Schumpeter recognized, an entrepreneur, an innovator, discovers new resources and new technological combinations, and change accelerates again, occasionally reaching the point at which it has the momentum to defeat the dominant realities of the large corporate structure or bureaucracy: capital, experience, organization. In this ecology of technology, evolution, not revolution, predominates; cycles proceed within cycles within cycles. The most salient factor is the rate of innovational change. And until we truly master the secrets of economic productivity and technological innovation—which we seem not to have accomplished so far—human effort and initiative will resemble a boy tossing a ball up in the air and watching it trace its distinctive parabolic arc against the sky.

# NOTES

## Chapter 1: The Gap between Science and Technology

1. John Lukacs, *Confessions of an Original Sinner* (New York: Ticknor and Fields, 1990), 131.
2. The full Oppenheimer quotation is: "Sometimes past knowledge is embodied not in a natural phenomenon but in an invention, or in elaborate pyramids of invention, a new technology." Robert Oppenheimer, *Atom and Void: Essays in Science and Community* (Princeton, N.J.: Princeton University Press, 1989), 17–18.
3. See Christopher Anderson, "Clinton's Technology Policy Emerges," *Science*, February 26, 1993, 1244; Eliot Marshall, "R&D Policy That Emphasizes the 'D,'" *Science*, March 26, 1993, 1816–19; and Christopher Anderson, "NSF Wins, NIH Loses in Clinton's 1994 Budget," *Science*, April 2, 1993, 24–25. On defense conversion of industry, see Eliot Marshall, "Swords to Plowshares Plan Boosts R&D," *Science*, March 19, 1993, 1690. For a look at managed trade and high technology, see Laura D'Andrea Tyson, *Who's Bashing Whom: Trade Conflicts in High Technology* (Washington, D.C.: Institute for International Economics, 1992).
4. There is a considerable literature on the relationship between innovation and corporate structure. For a general survey of the economic literature in the field, see Morton Kamien and Nancy Schwartz, *Market Structure and Innovation* (New York: Cambridge University Press, 1982).

## Chapter 2: The World of Tomorrow

1. H. G. Wells, "World of Tomorrow," *New York Times*, March 5, 1939, sec. 8, 4.
2. Many of the descriptions of the fair come from either the *New York*

*Time* throughout 1939 or Larry Zim, *The World of Tomorrow: The 1939 World's Fair* (New York: Harper & Row, 1988).

3. "A Glimpse behind the Scenes," *New York Times*, March 5, 1939, 21.

4. Alfred Chandler, Glenn Porter, and Harold Livesay, "The Structure of American Industry in the Twentieth Century: A Historical Overview," in *The Essential Alfred Chandler: Essays toward a Historical Theory of Big Business,* ed. Thomas McCraw (Boston: Harvard University Press, 1988). The description of prewar concentration could come from many sources, but the cited article focuses particularly on the structural fallout of the technically sophisticated chemical and electrical equipment industries.

5. Quote and statistics found in Theodore N. Beckman, "The Structure of Postwar American Business: Large versus Small Business after the War," *American Economic Review,* June 1944, 94.

6. National Science Foundation, *Federal Funds for Science* (Washington, D.C.: U.S. Government Printing Office, 1950).

7. Daniel Kevles, *The Physicists: The History of a Scientific Community in Modern America* (Boston: Harvard University Press, 1987).

8. Quoted in Kevles, *The Physicists,* 170.

9. Thomas Parran, Jr., "New Health for a New Age," *New York Times,* March 5, 1939, sec. 8, 43.

10. David Sarnoff, "Might of the Speeding World," *New York Times,* March 5, 1939, sec. 8, 15.

11. Arthur Compton, "The Goal of Science," *New York Times*, March 5, 1939, sec. 8, 11. Some of those predictions—notably, the emptying of urban areas into the suburbs—did come true.

12. "'Great Things Due,' Ford Says at Fair," *New York Times,* April 6, 1939, 22.

13. David Hounshell and John Kenly Smith, Jr., *Science and Corporate Strategy: DuPont R&D, 1902–1980* (New York: Cambridge University Press, 1988), 270.

14. Kenneth Bilby, *The General: David Sarnoff and the Rise of the Communications Industry* (New York: Harper & Row, 1986), 132–34.

15. For a more complete discussion of the development of certain science-based industries, see David F. Noble, *America by Design: Science, Technology and the Rise of Corporate Capitalism* (New York: Oxford University Press, 1977).

16. Chandler, Porter, and Livesay, "Structure of American Industry," 266.

17. Alfred Chandler, *Scale and Scope: The Dynamics of Industrial Capitalism* (Cambridge, Mass.: Belknap Press, 1990), 171.

18. John W. Servos, "The Industrial Relations of Science: Chemical Engineering at MIT, 1900–1939," *Isis,* December 1980, 531.

19. The discussion of Du Pont comes from Alfred Chandler, *The Visible Hand: The Managerial Revolution in American Business* (Cambridge, Mass.: The Belknap Press, 1977), 438–50; and Hounshell and Kenly

Smith, *Science and Corporate Strategy,* 223–48.

20. All quotes and description of Stine's reorganization of Du Pont R&D comes from Hounshell and Kenly Smith, *Science and Corporate Strategy,* 221–27.

21. For a complete exposition of the Manhattan Project, see Richard Rhodes, *The Making of the Atomic Bomb* (New York: Simon & Schuster, 1985).

22. Nuel Pharr Davis, *Lawrence and Oppenheimer* (New York: Da Capo Press, 1986), 196.

23. Rhodes, *Making of the Atomic Bomb,* 490–91.

24. Leslie Groves, *Now It Can Be Told: The Story of the Manhattan Project* (New York: Da Capo Press, 1962), 97.

25. Rhodes, *The Making of the Atomic Bomb,* 490.

26. Davis, *Lawrence and Oppenheimer,* 201–2.

27. The $2.5 billion figure comes from Hans Quiesser, *The Conquest of the Microchip: Science and Business in the Silicon Age* (Cambridge, Mass.: Harvard University Press, 1988). Calculating these figures is difficult because of the differences in the two projects. Radar was much more deeply integrated into the American commercial establishment. Hundreds of firms provided components and manufacturing, and thousands of different sets were produced. Thus, much of the money went for the mass production of hardware, which did not happen on the atom bomb project. The atom bomb was, in its own bloated way, a custom project, and thus more typical of postwar defense work; radar systems were built using mass-production techniques. Henry Guerlac, in *Radar in World War II* (American Institute of Physics, 1987, p. 3), estimates that some $3 billion worth of radar systems and associated equipment were delivered to the armed forces by 1945.

28. Quoted in John S. Rigden, *Rabi: Scientist and Citizen* (New York: Basic Books, 1987), 164.

29. "Alfred Lee Loomis: Amateur of the Sciences," *Fortune,* March 1946, 132. Alvarez characterization from Luis Alvarez, *Alvarez* (New York: Basic Books, 1987), 79.

30. David Fisher, *Race on the Edge of Time: Radar—The Decisive Leap of World War II* (New York: McGraw-Hill, 1988), 28.

31. L. A. du Bridge and L. N. Ridenour, "Expanded Horizons," *Technology Review,* May 1946, 23.

32. "Longhairs and Short Waves," *Fortune,* November 1945, 163.

33. Quoted in Rigden, *Rabi,* 118.

34. Du Bridge and Ridenour, "Expanded Horizons," 24.

35. Kevles, *The Physicists,* 302–3.

36. Quiesser, *Conquest of the Microchip,* 40.

## Chapter 3: Optimists and Pragmatists

1. Harley Kilgore, "Science and Government," *Science,* December 21, 1945, 630.

2. "The Scientists," *Fortune,* October 1948, 107.

3. On the Cowles Committee, see Herbert Simon, *Models of My Life* (New York: Basic Books, 1991), 101–7; and David Warsh, *Economic Principals: Masters and Mavericks of Modern Economics* (New York: Free Press, 1993), 64–70. Simon also spent time at Rand and was a pioneer in cybernetic theory. For a flavor of Cold Spring Harbor after the war, see Horace Freeland Judson, *The Eighth Day of Creation* (New York: Simon & Schuster, 1979). For the best study of the whiz kids, see John A. Byrne, *The Whiz Kids: The Founding Fathers of American Business—and the Legacy They Left Us* (New York: Currency/Doubleday, 1993). For the cybernetic meetings, see Steve J. Heims, *The Cybernetic Group* (Cambridge, Mass.: MIT Press, 1991).

4. For insight into the history of Raytheon, see Otto J. Scott, *The Creative Ordeal: The Story of Raytheon* (New York: Atheneum, 1974). All descriptions and statistics on early Raytheon history that follow come from Scott.

5. Vannevar Bush, *Pieces of the Action* (New York: William Morrow & Co., 1970).

6. Luis Alvarez, *Alvarez* (New York: Basic Books, 1987), 89.

7. Scott, *Creative Ordeal,* 163.

8. Information on Du Pont's nuclear effort comes from David Hounshell and John Kenly Smith, Jr., *Science and Corporate Strategy: DuPont R&D, 1902–1980* (New York: Cambridge University Press, 1988), 327–42.

9. Ibid., 336.

10. Ibid., 342.

11. Du Pont would finally be forced to sell its interest in General Motors in 1957.

12. For a discussion of Marshall's demise, see Scott, *Creative Ordeal,* 216–19.

13. "Economic Concentration and World War II," a report of the Smaller War Plants Corporation to the Special Committee to Study Problems of American Small Business, U.S. Senate, 79th Congress, 2nd Session, June 14, 1946, 29.

14. Theodore Beckman, "The Structure of Postwar American Business: Large versus Small Business after the War," *American Economic Review,* June 1944, 94.

15. I. F. Stone, "The Cartel Cancer," *The Nation,* February 12, 1944, 178; and Stone, "Cartel's Washington Friends," *The Nation,* February 19, 1944, 210.

16. "People in the Limelight," *The New Republic,* January 8, 1945, 40.

17. Daniel Kevles, *The Physicists: The History of a Scientific Community in Modern America* (Boston: Harvard University Press, 1987), is illuminating on both Kilgore and Vannevar Bush. Also see Robert Franklin Maddox, *The Senatorial Career of Harley Kilgore* (New York: Garland Publishing, 1981).

18. The dispute between Kilgore and Bush is examined from different perspectives by Kevles in *The Physicists* and by Nathan Reingold in *Science, American Style* (New Brunswick, N.J.: Rutgers University Press, 1991).

19. "The Great Science Debate," *Fortune*, January, 1946, 236.

20. Reingold, *Science, American Style*, 287.

21. John W. Servos, "The Industrial Relations of Science: Chemical Engineering at MIT, 1900–1939," *Isis*, December 1980, 531.

22. Daniel Kevles, "FDR's Science Policy," *Science*, March 1, 1974, 798–800.

23. Vannevar Bush, "The Kilgore Bill," *Science*, December 31, 1943, 571.

24. Maddox, *Senatorial Career of Harley Kilgore*, 172.

25. Ibid., 171.

26. Reingold, *Science, American Style*, 321.

27. "The Limitations of Science," *Time*, May 7, 1963, 81.

## Chapter 4: The Conservatism of Television

1. "RCA's Television," *Fortune*, September 1948, 80.

2. Ibid., 81.

3. "Unprecedented Growth in Industrial History, Says Frank Folsom," *Broadcasting and Telecasting*, January 2, 1950, 3.

4. *New York Times*, "U.S. Industry Start Expected with N.Y. World's Fair Opening Telecasts," April 6, 1939, 22.

5. For a general overview of Sarnoff, RCA, and the history of radio, see Tom Lewis, *Empire of the Air: The Men Who Made Radio* (New York: HarperCollins, 1991). The most complete biography of Sarnoff is Kenneth Bilby, *The General: David Sarnoff and the Rise of the Communications Industry* (New York: Harper & Row, 1986). Robert Sobel, *RCA* (New York: Stein & Day, 1986), goes into the greatest detail about the company. For a history of the early years of television, particularly in terms of regulation, see Erik Barnouw, *The Golden Web: A History of Broadcasting in the United States, 1933–1953* (New York: Oxford University Press, 1968).

6. "Commander McDonald of Zenith," *Fortune*, June 1945, 141.

7. Erik Barnouw, *Tube of Plenty: The Evolution of American Television* (New York: Oxford University Press, 1990), 9–12.

8. Quoted in Frank C. Waldrop and Joseph Borkin, *Television: A Struggle for Power* (New York: William Morrow & Co., 1938), 223–24.

9. "RCA's Television," 81–82.

10. For RCA's manufacturing problems and the hiring of Folsom, see "RCA's Television," 82–83; and William B. Harris, "RCA Organizes for Profit," *Fortune,* August 1957, 110.

11. Sobel's *RCA* (pp. 150–67) provides volume figures on early television sales, a description of early models (including the "Model T"), and an analysis of RCA's difficulties.

12. On Philco, see "Radio, Refrigerators and Radar," *Fortune,* November 1945, 115.

13. "In Television, Admiral's Hot," *Fortune,* June 1949, 89.

14. Ibid., 90.

15. Ibid., 126.

16. "Admiral Sells Television through Television," *Broadcasting and Telecasting,* January 23, 1950, 64.

17. Sobel, *RCA,* 151.

18. Scott, *Creative Ordeal,* 247–68.

19. Sobel, *RCA,* 166.

20. "Color TV Makes the Scene Again," *Business Week,* September 16, 1967, 153; and Harvard Business School, *Zenith Radio Corp.* (Boston: HBS Case Services, 1983).

21. Sobel, *RCA,* 166.

22. Edward T. Thompson, "The Upheaval at Philco," *Fortune,* February 1959, 113.

23. "Ford's Big Deal to Expand Space Work," *Business Week,* September 26, 1961, 61; and "Ford-Philco Merger," *Financial World,* September 27, 1961, 1.

24. Harvard Business School, *The U.S. Television Set Market, 1970–1979* (Boston: HBS Case Services, 1982); and Harvard Business School, *Zenith and the Color Television Fight* (Boston: HBS Case Services, 1982).

25. John E. Henneberger, "Productivity Rises as Radio-TV Output Triples in Eight Years," *Monthly Labor Statistics,* March 1969, 40.

26. E. K. Faltermayer, "The Coming Battle for the Color TV Market," *Fortune,* January 1966, 190; and Harvard Business School, *Zenith Radio Corp.* At the same time that the dispute over circuit boards was occurring, RCA was considering using more transistors. "I don't know of anything before an integrated circuit that could make a substantial difference in the kind of set the industry puts out," *Fortune* quoted Sarnoff. But another RCA executive made a candid, very un-RCA-like admission: "We're gambling that the competition won't come out with a transistorized set in the meantime." The Japanese already had. Sobel, in *RCA* (pp. 210–14), blames RCA under Sarnoff for failing to move into the Japanese market and licensing away much of the technology to the Japanese.

27. Harvard Business School, *U.S. Television Set Market,* 31.

28. Quoted in, ". . . Meanwhile, Back in the Limelight," *Forbes,* March 15, 1968, 89.

29. "Playing It Straight," *Forbes*, November 1, 1972, 64.
30. "Here's a Rome Arrangement...," *Chicago Tribune*, September 10, 1973, sec. 2, p. 14.
31. Leonard Wiener, "New Owner Could Sharpen View of Admiral's Future," *Chicago Tribune*, October 28, 1974, sec. 2, p. 9.
32. Harvard Business School, *U.S. Television Set Market, 1970–1979*, 23–24. "In the late 1960s the Japanese were by and large regarded as lagging their U.S. and European competitors in the areas of set design and labor productivity. By the late 1970s, the situation had reversed" (p. 23).
33. *Consumer Reports Buying Guide*, 1968–80. The comments on repairing demo models comes from the 1974 guide, p. 3.
34. On the last days of Sarnoff, see Sobel, *RCA*, 207–8; and Lewis, *Empire of the Air*.
35. Sobel, *RCA*, 207.
36. The literature on HDTV is quite large and growing. For different perspectives, see Harvard Business School, *Zenith and High-Definition Television* (Boston: HBS Case Studies, 1991); and G. Donlan, *Supertech: How America Can Win the Technology Race* (Homewood, Ill.: Business One Irwin, 1991), 1–41. For a broader argument about new uses for television and technology policy, see George Gilder, *Life After Television: The Coming Transformation of Media and American Life* (New York: W.W. Norton & Co., 1992).

## Chapter 5: Centrifugal Tendencies

1. Frank Dunstone Graham, *Social Goals and Economic Institutions* (Princeton, N.J.: Princeton University Press, 1942), 211.
2. I. F. Stone, "The Cartel Cancer," *The Nation*, February 12, 1944, 198.
3. Joseph Schumpeter, *Capitalism, Socialism and Democracy* (New York: Harper Torchbooks, 1975), 100–101.
4. Ibid., 106. The full quote is: "[T]he perfectly competitive arrangement displays waste of its own. The firm of this type that is compatible with perfect competition is in many cases inferior in internal, especially technological efficiency.... And as we have seen before, a perfectly competitive industry is much more apt to be routed—and to scatter the bacilli of depression—under the impact of progress or of external disturbance than is big business."
5. W. Rupert McClaurin, *Invention and Innovation in the Radio Industry* (New York: Macmillan, 1949), 249.
6. There are a number of sources on the early history of AT&T, including John Brooks, *Telephone: The Wondrous Invention That Changed the World and Spawned a Corporate Giant* (New York: Harper & Row, 1975); and Thomas Hughes, *American Genesis* (New York: Viking Press, 1989).

7. Hughes, *American Genesis*, 157.

8. Brooks, *Telephone*, 205–6.

9. For a description of Bell Labs in the 1930s, and a close analysis of its development, see Leonard Reich, *The Making of American Industrial Research: Science and Business at GE and Bell* (New York: Cambridge University Press, 1985), 204.

10. Quoted by Nathan Reingold, *Science, American Style* (New Brunswick, N.J.: Rutgers University Press, 1991), 301–2.

11. See Daniel Kevles, *The Physicists: The History of a Scientific Community in Modern America* (Boston: Harvard University Press, 1987).

12. Frank B. Jewett and Ralph W. King, "Engineering Progress and Social Order," speech given at the University of Pennsylvania Bicentennial Celebration, 1941.

13. "Jewett Opposes Science Agency," *New York Times*, October 15, 1946, 3.

14. Charles Hessian, *Galbraith and His Critics* (New York: New American Library, 1972), 25. In his memoirs, *A Life in Our Times* (Boston: Houghton Mifflin, 1981, p. 175), Galbraith puts it slightly differently: "I decided that henceforth I would submit myself to a wider audience, a decision that, in contrast with some others, I have not regretted."

15. John Kenneth Galbraith, *American Capitalism: The Concept of Countervailing Power* (Boston: Houghton Mifflin, 1952), 91. Galbraith believed far more in oligopoly than he did in monopoly. "While it may be going too far to say that oligopoly insures progress, technical development is all but certain to be one of the instruments of commercial rivalry when the number of firms is small," he wrote (p. 94). ". . . There is reason to suppose that an industry characterized by oligopoly will be more progressive than an industry controlled by monopoly" (p. 151).

16. Galbraith, *American Capitalism*, 91.

17. All Weissman quotes are from Rudolph Weissman, *Small Business and Venture Capital* (New York: Harper & Row, 1945), 1–3.

18. Keynes quote from John Maynard Keynes, *The General Theory of Employment, Interest and Money* (New York: Harcourt Brace Jovanovich, 1953), 159.

19. Weissman, *Small Business*, 58.

20. An overview of the condition of small business immediately after the war can be found in Thomas F. Murphy, "The Big Worry for Small Business: Money," *Fortune*, July 1957, 120.

21. Ibid., 121.

22. Peter Drucker, "How Big Is Too Big?" *Harper's*, July 1950, 23–28; and Peter Drucker, "The Care and Feeding of Small Business," *Harper's*, August 1950, 74–79.

23. Cole was also part of a large movement in business history that eventually produced Alfred Chandler and his many disciples. For a short treatment on this subject, see David Warsh, "Of Kings, Cabbages and Robert Reich," in his collection *Economic Principals: Masters and Mavericks of Modern Economics* (New York: Free Press, 1993), 470–73.

24. "War, Cash and Corporations," *Fortune*, April 1945, 114; and "The Boom," *Fortune*, June 1946, 97.

25. "Where Do Small Firms Get Capital?" *Business Week*, December 11, 1948, 74–75.

26. Michael Bernstein, *The Great Depression: Delayed Recovery and Economic Change in America, 1929–1939* (New York: Cambridge University Press, 1987).

27. Quoted in *New York Times* obituary, February 2, 1982, sec. B, p. 16.

28. Details of Whitney's life come from E. J. Kahn, *Jock: The Life and Times of John Hay Whitney* (Garden City, N.Y.: Doubleday & Co., 1981).

29. Kahn, *Jock*, 184–90.

30. The evolution of Flanders's views is traced by Edwin Layton, *The Revolt of the Engineers: Social Responsibility and the American Engineering Profession* (Baltimore, Md.: Johns Hopkins University Press, 1986), 232–33. Galbraith discusses writing "Toward Full Employment" in *A Life in Our Times*, 66.

31. Ralph Flanders, *Senator from Vermont* (Boston: Little, Brown & Co., 1961), 186–89. Also see "Flanders of New England," *Fortune*, August 1945, 135.

32. Quoted by William D. Bygrave and Jeffrey Timmons, *Venture Capital at the Crossroads* (Boston: Harvard Business School Press, 1992), 17. The original source is an unpublished Harvard dissertation on ARD, "Sustaining the Venture Capital Firm," by Patrick Liles.

33. Georges Doriot and Cecil Eaton Fraser, *Analyzing Our Industries* (New York: McGraw-Hill, 1932).

34. *New York Times*, June 11, 1933, sec. 2, p. 1.

35. Gene Bylinsky, "General Doriot's Dream Factory," *Fortune*, August 1967, 103.

36. Quoted in Doriot obituary, *Boston Globe*, June 3, 1987, p. 59.

37. Lawrence Shames, *The Big Time—Harvard Business School's Most Successful Class and How It Shaped America* (New York: Harper & Row, 1986), 107.

38. Bylinsky, "General Doriot's Dream Factory," 104.

39. Schumpeter, *Capitalism, Socialism and Democracy*, 132.

40. John M. Blair, "Does Large-Scale Enterprise Result in Lower Costs? Technology and Size," *American Economic Review*, May 1948, 121–53.

41. Ibid.

## Chapter 6: The Entrepreneurial Transistor

1. The invention of the transistor is described in John Brooks, *Telephone: The Wondrous Invention That Changed the World and Spawned a Corporate Giant* (New York: Harper & Row, 1975); Richard Nelson, "The Link between Science and Invention: The Case of the Transistor," in *Rate and Direction of Inventive Activity: Economic and Social Factors* (Princeton, N.J.: Princeton University Press for the National Bureau for Economic Research, 1962), 549–81; and Ernest Braun and Stuart Macdonald, *Revolution in Miniature: The History and Impact of Semiconductor Electronics* (New York: Cambridge University Press, 1982), 33–44.

2. Braun and Macdonald, *Revolution in Miniature*, 38.

3. Ibid. See also John E. Tilton, *International Diffusion of Technology* (Washington, D.C.: Brookings Institution, 1971).

4. Figures come from National Science Foundation, *Federal Funds for Science* (Washington, D.C.: U.S. Government Printing Office, 1950–51, 1951–52, 1953–54, 1954–55, 1955–56, 1956–57).

5. On the plight of higher education in the early 1950s, see Seymour Harris, "The Threefold Crisis in Education," *New York Times Magazine*, October 30, 1949, p. 16; "Colleges in Trouble," *Business Week*, January 20, 1951, 100; and "Colleges Face a Private Depression," *Business Week*, April 7, 1951, 52.

6. On Frederick Terman, see Dirk Hanson, *The New Alchemists: Silicon Valley and the Microelectronics Revolution* (Boston: Little, Brown, 1982), 86–88; Stuart Leslie, *The Cold War and American Science* (New York: Columbia University Press, 1993); and Roger L. Geiger, "Science, Universities and National Defense, 1945–1970," *Osiris*, vol. 7, 1992, 26–48.

7. On Stanford's real estate efforts, see "Industry Blooms on Campus," *Business Week*, February 17, 1951, 94; and "Land-Poor Stanford Opens Its Acres in 99-Year Lease," *Business Week*, December 20, 1952, 134.

8. On the early military role in transistor development, see Braun and Macdonald, *Revolution in Miniature*, 70–72.

9. Braun and Macdonald, *Revolution in Miniature*, 145.

10. Ibid., 48–49.

11. For an overview of the competitive situation in the early 1950s, see Lawrence P. Lessing, "The Electronics Era," *Fortune*, July 1951, 79. For a description of the battle with the vacuum tube makers, including data on innovations and federal funding, see Braun and Macdonald, *Revolution in Miniature*, 59–61; Tilton, *International Diffusion of Technology*, 49–97; Karl S. Elebash, Jr., "Durable Transistors to Oust More Tubes in Electronic Gear," *Wall Street Journal*, May 23, 1955, p. 1; and William B. Harris, "The Battle of the Components," *Fortune*, May 1957, 135.

12. The best portrait of Texas Instruments in the early days is given in a two-part story by John McDonald, "The Men Who Made T.I.," *Fortune*, November 1961, 117; and "Where Texas Instruments Goes from Here," *Fortune*, December 1961, 110.

13. Patrick Haggerty, "Strategies, Tactics and Research," *Research Management*, 1966, 148.

14. The following discussion on finance is based on Haggerty's *Research Management* article and on figures from *Moody's Industrial Guide* for the relevant years.

15. Quoted in Haggerty, "Strategies, Tactics and Research," 151.

16. General background on Transitron is from William B. Harris, "The Company That Started with a Gold Whisker," *Fortune*, August 1959, 98.

17. "Transitron Sets Investors Agog," *Business Week*, December 5, 1959, 123.

18. On the early 1960s transistor slump, see Charles E. Silberman, "The Coming Shakeout in Electronics," *Fortune*, August 1960, 126; McDonald, "Where Texas Instruments Goes from Here," 110; and Braun and Macdonald, *Revolution in Miniature*, 81–87.

19. The later history of Transitron is recounted in Ellyn Spragins, "Remember Transitron?" *Forbes*, June 6, 1983, 200.

20. An account of Shockley can be found in Robert Slater, *Portraits in Silicon* (Cambridge, Mass.: MIT Press, 1987).

21. Shockley's Raytheon negotiations are mentioned in Braun and Macdonald, *Revolution in Miniature*, 125. The authors refer back to an article describing the Shockley-Fairchild family tree, "Semiconductor Family Tree," *Electronic News*, July 8, 1968, 4–5, 38.

22. Life at Shockley is recounted in Dirk Hanson, *The New Alchemists: Silicon Valley and the Microelectronics Revolution* (Boston: Little, Brown, 1982), 88–92.

23. Quote from Eugene Kleiner in John Wilson, *The New Venturers: Inside the High-Stakes World of Venture Capital* (Reading, Mass.: Addison-Wesley, 1985), 32.

24. Beckman annual report quotes from Harrison Stephens, *Golden Past, Golden Future: The First Fifty Years of Beckman Instruments* (Claremont, Calif.: Claremont University Center, 1985), 72.

## Chapter 7: The Ethos of the Market

1. S. Devlin, "Hopeful Signs vs. Troubled Spots for Electronic and TV Companies," *Magazine of Wall Street*, August 2, 1958, 522.

2. "The Battle for Investment Survival," *Financial World*, January 28, 1959, 19.

3. Benjamin Graham, David Dodd, and Sidney Cottle, *Security Analysis: Principles and Techniques* (New York: McGraw-Hill, 1962), 413–15.

4.  Ibid., 426–31.

5.  T. Rowe Price, "Picking Growth Stocks in the 1950s," in *Classics II: Another Investor's Anthology* (Homewood, Ill.: Business One Irwin, 1991), 383. The article originally appeared in two issues of *Barron's*: February 6, 1950 (p. 13), and February 20, 1950 (p. 19).

6.  Gilbert Burck, "A New Kind of Stock Market," *Fortune*, March 1959, 120. Also see "New Surge to Growth Stocks," *Business Week*, September 27, 1958, 31.

7.  Graham, Dodd, and Cottle, *Security Analysis*, 429. Footnote on p. 429 elaborates on the Stanford Research Institute study that correlates R&D expenditures with profits.

8.  On Doriot and ARD, see William D. Bygrave and Jeffrey A. Timmons, *Venture Capital at the Crossroads* (Boston: Harvard Business School Press, 1992), 18–21. Doriot was on the cover of *Business Week* for February 19, 1949.

9.  On Doriot and DEC, see Bygrave and Timmons, *Venture Capital*, 20; Kenneth H. Olsen, "Digital Equipment Corp., the First Twenty-five Years," The Newcomer Society in North America, 1983; and Glenn Rifkin, *The Story of Ken Olsen and the Digital Equipment Corporation* (Chicago: Contemporary Books, 1988).

10. See Walter A. McDougall, . . . *The Heavens and the Earth: A Political History of the Space Age* (New York: Basic Books, 1985), 41—56.

11. For further information on the Small Business Investment Corporations, see Bygrave and Timmons, *Venture Capital*, 21–23. Also see "A Bank for Small Business," *Business Week*, April 26, 1958, 148; and "SBI Show Is on the Road," *Business Week*, December 6, 1958, 28. Clifford L. Fitzgerald, Jr., reviewed the business reaction to the SBICs in "Small Business Financing," *Harvard Business Review*, April 1959, 6. Several years later, Samuel Hayes evaluated their effectiveness, in "Are SBICs Doing Their Job?" *Harvard Business Review*, March–April 1963, 7.

12. See Leonard Silk, *The Research Revolution* (New York: McGraw-Hill, 1960); Kenneth Boulding, *The Meaning of the Twentieth Century: The Great Transition* (New York: Harper & Row, 1964), on the concept of postcivilization; Sumner Slichter, *Economic Growth in the United States: Its History, Problems and Prospects* (New York: Collier Books, 1963), on the three phases. Quote on diversity producing stability comes from Slichter.

13. "Dynamic Electronics Industry," *Magazine of Wall Street*, March 26, 1960, 24.

14. Arthur Merrill, *Investing in the Scientific Revolution* (New York: Doubleday & Co., 1962).

15. Graham, Dodd, and Cottle, *Security Analysis*, 57.

16. Merrill, *Investing in the Scientific Revolution*, 1.

17. Ibid., 2.

18. Ibid., 4.
19. Ibid., 9.
20. "The Egghead Millionaires," *Fortune*, September 1960, 172.
21. For a review of Kuznets's book, see Peter Bernstein, "Capital in the American Economy," *Harvard Business Review*, March–April 1962, 170.
22. Benjamin Graham, *The Intelligent Investor* (New York: Harper & Row, 1972), 34.
23. See Robert Sobel, *The Last Bull Market: Wall Street in the 1960s* (New York: Norton, 1980), particularly for a sense of the market gyrations. See also John Brooks, *The Go-Go Years* (New York: Weybright and Talley, 1973); and Adam Smith, *Money Game* (New York: Random House, 1968).
24. Quoted in Robert Sobel, *The Rise and Fall of the Conglomerate Kings* (New York: Stein and Day, 1984), 98. The quote originally appeared in John Wall, "Want to Get Rich Quick? An Expert Gives Some Friendly Advice on Conglomerates," *Barron's*, February 5, 1968), 19.
25. Robert J. Schoenberg, *Geneen* (New York: Warner Books, 1985), 138–39.
26. Buckminster Fuller, *Synergetics* (New York: Macmillan, 1975). Fuller elaborates on synergy, pp. 3–13; on synergetics, pp. 22–67.
27. On the roots of the word synergy, see "synergy" in *The Oxford English Dictionary*, second edition, vol. 17, 481 (Oxford: Clarendon Press, 1989). On the concept in science, see "2 + 2 = 5," *Scientific American*, September 1950, 46. On Tex Thornton and synergy, see John A. Byrne, *The Whiz Kids: The Founding Fathers of American Business— and the Legacy They Left Us* (New York: Currency/Doubleday, 1993), 180.
28. "Textron: Yankee-Style Conglomerate," *Financial World*, July 9, 1969, 5.
29. Quoted in Chris Welles, "Venture Capital: The Biggest Mousetrap of the 1970's?" *Institutional Investor*, January 1970, 40.
30. Ibid., 44.
31. Peter L. Bernstein, *Capital Ideas: The Improbable Origins of Modern Wall Street* (New York: Free Press, 1992).

## Chapter 8: The Balance of Microelectronic Power

1. The Kleiner/Rock anecdote is from John Wilson, *The New Venturers: Inside the High-Stakes World of Venture Capital* (Reading, Mass.: Addison-Wesley, 1985).
2. Quoted in Wilson, *New Venturers*, 34.
3. Quoted in Dirk Hanson, *The New Alchemists: Silicon Valley and the Microelectronics Revolution* (Boston: Little, Brown, 1982), 93.

4.  The Fairchild family tree is described in Wilson, *New Venturers;* Hanson, *New Alchemists;* and Ernest Braun and Start Macdonald, *Revolution in Miniature: The History and Impact of Semiconductor Electronics* (New York: Cambridge University Press, 1982).

5.  See Wilson, *New Venturers,* particularly 31–48.

6.  "Chronic entrepreneurialism" appears in Martin Kenney and Richard Florida, *The Breakthrough Illusion: Corporate America's Failure to Move from Innovation to Mass Production* (New York: Basic Books, 1990). Wilson, *New Venturers,* uses the term "vulture capitalist."

7.  Wilson, *New Venturers,* 194.

8.  Quoted by Larry Hicks in "Empty Promises in Big Chip Deal?" *Sacramento Bee,* November 5, 1989. Rodgers was responding to the formation of a U.S. R&D consortium, U.S. Memories. "It's one of the dumbest things I've ever heard," said Rodgers. "Forming the Amtrak of the semiconductor business is no way competitive with the Japanese."

9.  Martin Kenney and Richard Florida criticize the market-oriented entrepreneurialism of Silicon Valley and argue for a more Japanese (or German) system of large corporate agglomerations protected from the financial marketplace and nurtured by the government. Going further, they also argue that the Japanese manufacturing systems offer the same kind of corporatist ideal—the destruction of hierarchy, the elimination of conflict between worker and manager—that Peter Drucker has long talked about (Kenney and Florida, *Breakthrough Illusion;* and Martin Kenney and Richard Florida, *Beyond Mass Production: The Japanese System and Its Transfer to the U.S.* (New York: Oxford University Press, 1993). Charles H. Ferguson supports the Kenney/Florida position in "Computers and the Coming U.S. Keiretsu," *Harvard Business Review,* July–August 1990, 55.

10. For an overview of the evolution of the American semiconductor industry, see Braun and Macdonald, *Revolution in Miniature;* and John Tilton, *International Diffusion of Technology* (Washington D.C.: Brookings Institution, 1971).

11. William D. Bygrave and Jeffrey A. Timmons, *Venture Capital at the Crossroads* (Boston: Harvard Business School Press, 1992), 25–30.

12. "Ibbotson and Sinquefield View the Future," *Institutional Investor,* July 1992, 54.

13. For statistics on growth and venture capital, see Bygrave and Timmons, *Venture Capital,* 27. For pension fund figures, see 45–46.

14. Harvard Business School, *LSI Logic Corporation* (Boston: HBS Case Services, 1990).

15. Author interview with industry observer Richard Shaffer, editor, *Computer Letter,* February 12, 1988.

16. The term *virtual reality company* was popularized by William H.

Davidow and Michael S. Malone, *The Virtual Corporation* (New York: Edward Burlingame Books/Harper Business, 1992). Also see John Byrne, "The Virtual Corporation," *Business Week,* February 8, 1993, 98.

17.  See Harvard Business School, *The Global Semiconductor Industry, 1987* (Boston: HBS Case Services, 1991); and Laura D'Andrea Tyson, *Who's Bashing Whom: Trade Conflicts in High-Technology* (Washington, D.C.: Institute for International Economics, 1992), 85–154.

18.  Harvard Business School, *Intel Corporation 1988* (Boston: HBS Case Services, 1991).

19.  For industry economics, see Braun and Macdonald, *Revolution in Miniature;* Tilton, *International Diffusion of Technology;* and Harvard Business School, *Global Semiconductor Industry.* For the later impact of semiconductors on the computer industry, see Charles H. Ferguson and Charles R. Morris, *Computer Wars: How the West Can Win in a Post-IBM World* (New York: Times Books, 1993).

20.  While there was only one synchrotron used for etching chips with X rays in the United States at the end of the 1980s, there were ten such devices in Japan. See Ferguson and Morris, *Computer Wars,* 119.

21.  George Gilder, *Microcosm* (New York: Simon & Schuster, 1989).

22.  On the shifting balance among Intel, Microsoft, and IBM, see Ferguson and Morris, *Computer Wars.* Also see Thomas Mccarroll, "IBM's Unruly Kids," *Time,* February 1, 1993.

23.  Ferguson and Morris, *Computer Wars,* 56–81. The authors delve into the failure of IBM to pursue an IBM invention—the reduced instruction set computer (RISC) chip. RISC was another of the software-intensive techniques that powerfully increased chip speeds and changed the dynamics of the business in the 1980s. IBM's commercial efforts in RISC technology were belated and weak, while competitors such as Sun Microsystems swept past it in the workstation market.

24.  For details on Microsoft and its battles with IBM, see Ferguson and Morris, *Computer Wars,* 82–97.

25.  For further discussion of the Microsoft–IBM divorce and the development of Windows and Windows NT, see Ferguson and Morris, *Computer Wars,* 90–97.

26.  On Intel, Microsoft, and their antitrust problems, see "Microsoft: Is It Too Powerful?" *Business Week,* March 1, 1993, 82; "The Politics of Chips," *Business Week,* March 1, 1993, 86; and "Little Guys Who Don't Scare," *Business Week,* March 1, 1993, 88.

27.  Author interview with industry observer Richard Shaffer, March 25, 1993. Also, on Intel's strategic maneuvering, see "Follow the Leader," *Computer Letter,* August 20, 1990; and "Fine Tuning," *Computer Letter,* August 26, 1991. On the increasing importance of manufacturing and distribution, see "PC Economics 101," *Computer Letter,* Novem-

ber 18, 1991. On what it will require for software competitors to keep pace with Microsoft, see "Open and Shut Case," *Computer Letter,* February 1, 1993.

28. Ferguson and Morris, *Computer Wars,* 139–40.
29. Ibid., 127.

## Chapter 9: Betting on Drug Discovery

1. Lewis Thomas, *The Youngest Science: Notes of a Medicine Watcher* (New York: Basic Books, 1983), 29.
2. Ibid., 28, 35.
3. See "Vannevar Bush's New Deal for Research," in Nathan Reingold, *Science, American Style* (New Brunswick, N.J.: Rutgers University Press, 1991).
4. On the prewar drug industry, see Tom Mahoney, *The Merchants of Life: An Account of the American Pharmaceutical Industry* (New York: Harpers, 1959).
5. Quoted in ibid.
6. For a perspective on the evolution of the industry, see E. W. Axe & Co., *The Drug Industry* (New York: E. W. Axe & Co., 1953), 9–12.
7. "Pfizer's Self-Prescribed Tonic," *Fortune,* August 1965, 153.
8. For pricing data, see E. W. Axe & Co., *The Drug Industry,* 13–18. On new drug introductions, see Wyndham Davies, *The Pharmaceutical Industry* (London: Pergamon Press, 1966), 7.
9. Axe & Co., *Drug Industry,* 16.
10. Charles L. Fontenay, *Estes Kefauver: A Biography* (Knoxville: University of Tennessee Press, 1980). Comment on Blair appears on p. 112.
11. Ibid. For further perspectives on the Kefauver drug hearings, see also Joseph Bruce Gorman, *Kefauver: A Political Biography* (New York: Oxford University Press, 1971); and Davies, *Pharmaceutical Industry.* For a survey of the debate over administered prices, see "Are Some Key Industries Pushing Up Inflation?" *Business Week,* October 6, 1975, 46.
12. Quoted in Fontenay, *Estes Kefauver,* 381.
13. the price-fixing scandal, see "FTC Rules on Tetracycline Case," *Business Week,* December 10, 1966, 46; and "Drugs: An Epidemic of Law Suits?" *Fortune,* February 1, 1968, 35.
14. For a somewhat feverish description of the Russell Long speech, see Ed Cray, *The Pill Pushers* (North Hollywood, Calif.: Brandon House, 1966), 5–25.
15. For pricing data, see David Schwartzman, *Innovation in the Pharmaceutical Industry* (Baltimore, Md.: Johns Hopkins University Press, 1976), 274–78. Schwartzman discusses pricing of tetracyclines without mentioning either the effect of the Kefauver hearings or the FTC charges. The five companies were eventually convicted of price fixing.

See "Drugs: An Epidemic of Law Suits?" *Forbes,* February 1, 1968, 35.

16. On the neglect of the lone inventor, see John Blair, *Economic Concentration: Structure, Behavior and Public Policy* (New York: Harcourt Brace Jovanovich, 1972), 238–41 and chap. 9, "Invention and Innovation"—notably, his discussion of wonder drugs, p. 216.

17. Davies, *Pharmaceutical Industry,* 32. Davies continues: "Critics of the industry often forget that however intelligently a research programme is worked out, there is no more certainty that it will yield profitable results than there is that studying form will infallibly indicate a winning horse," 32.

18. Schwartzman, *Innovation in the Pharmaceutical Industry,* 31.

19. Ibid. Schwartzman explicitly argues the Schumpeterian case for large drug companies, pp. 62–65.

20. For quotes on Schwartzman's hypothesis, see "The Hidden Cost of Drug Safety," *Business Week,* February 21, 1977, 80.

21. Ibid.

22. For an overview of the industry in the mid-1960s, see Davies, *Pharmaceutical Industry.*

23. "Hidden Cost of Drug Safety," 80.

24. On the cortisone rage, see Leonard Engel, "ACTH, Cortisone & Co.," *Harper's,* August 1951, 25; Leonard Engel, "Cortisone and Plenty of It," *Harper's,* September 1951, 56; Bruce Bliven, "News of the Miracle Hormones," *New Republic,* August 25, 1951, 14; "New Hope for Cortisone for All," *Business Week,* July 14, 1951, 26; and Milton Silverman, "The Wonderful Medicine Plant," *Saturday Evening Post,* February 21, 1953, 28.

25. "The Cortisone Shortage," *Fortune,* May 1953, 83.

26. "Synthesis of a Steroid," *Scientific American,* June 1951, 30.

27. See *A Corporation and a Molecule: The Story of Research at Syntex* (Palo Alto, Calif.: Syntex, 1966); Carl Djerassi, *The Pill, Pygmy Chimps, and Degas' Horse* (New York: Basic Books, 1992); and "Mexican Hormones," *Fortune,* May 1953, 88.

28. "Cortisone and Plenty of It," *Harper's,* September 1951, 56.

29. Djerassi, *The Pill,* 27.

30. Ibid., 50. The Djerassi autobiography is full of details about commercial arrangements at Syntex.

31. The two best profiles of Charlie Allen are Robert Sheehan, "Charlie Allen: A Feel for Stocks," *Fortune,* May 1954, 124; and Stanley Penn, "Little Known Financier, Astute but Unorthodox, Builds a Huge Fortune," *Wall Street Journal,* August 4, 1979, 1.

32. *Business Week* wrote about Syntex as a regulatory market issue in "Watching Brushfires on Wall Street," January 25, 1964, 110. William M. Carley focused on Syntex rumors in "Syntex's History Discloses Many Rumors, Latest of Which Sent Stock Price Down," *Wall Street*

*Journal,* August 26, 1964. The *New York Herald Tribune* captured the roller-coaster ride of Syntex on Wall Street in "The Legendary Syntex—From Rags to Riches," January 19, 1964, sec. 3, 2. *Forbes* captured the downside of its ride in "They Had a Dream," November 15, 1968, 29; and the operating realities after the go-go years were discussed in "Company with a Split Personality," *Forbes,* November 1, 1973, 64.

33.   For an overview of the scale economics of the global drug business, see Robert Teitelman, "Global Report on Pharmaceuticals," *Financial World,* May 30, 1989, 54; and "Sadie Hawkins Day," *Financial World,* September 5, 1989.

34.   For a description of Glaxo and its Zantac campaign, see Robert Teitelman, "Staying Power," *Financial World,* April 4, 1989, 28.

## Chapter 10: Biotechnology's Incomplete Revolution

1.   On the founding of Syntex's various Palo Alto–based spinoffs and its involvement with molecular biology, see Carl Djerassi, *The Pill, Pygmy Chimps, and Degas' Horse* (New York: Basic Books, 1992).

2.   On Syntex research productivity in the 1980s, see Robert Teitelman and Anthony Baldo, "Grading R&D," *Financial World,* January 24, 1989, 22. Syntex came in last among the major pharmaceutical companies. "Ironically, since its days as the birth control pill company, Syntex has had a reputation as a good research house. And it does spend a higher percentage of sales on R&D than any other company, 17 percent in 1988. Alas, despite a healthy 20 percent earnings growth over 10 years, Syntex only generated a bit over $200 million in new sales throughout the decade. . . . Syntex's R&D productivity problem bolsters an argument made by Squibb's Sanders. 'In order to play in the major leagues you need a critical mass of money,' he says. 'I'm not sure what that level is—but it's well in excess of $200 million a year.' Syntex only broke that barrier last year."

3.   "Pfizer's Self-Prescribed Tonic," *Fortune,* August 1965, 153. By the mid-1960s even the established drug companies recognized the potential in molecular biology. As *Fortune* wrote: "In the last few years, drug research has been moving from a pragmatic hit-or-miss search for chemicals with observable therapeutic effects into a pursuit aimed at unraveling the basic mysteries of human chemistry. . . . Work on the nucleic acids, for example, may lead to drugs that will combat viruses and perhaps even have a chilling power to influence memory, aging and heredity." *Fortune* quoted one Pfizer researcher: "The name of the game [over the long term] is molecular biology."

4.   Quoted in Horace Freeland Judson, *The Eighth Day of Creation* (New York: Simon & Schuster, 1979), 72. For a more detailed study of

Weaver see Lily Kay, *The Molecular Vision of Life* (New York: Cambridge University Press, 1993).

5. See Ernst Mayr, *A History of Biological Thought: Diversity, Evolution and Inheritance* (Cambridge, Mass.: Harvard University Press, 1982), 808–28.

6. Judson, *Eighth Day of Creation*, 201.

7. See Joseph S. Fruton, *Molecules and Life: Historical Essays on the Interplay of Chemistry and Biology* (New York: Wiley, 1972); Joseph S. Fruton, *A Skeptical Biochemist* (Cambridge, Mass.: Harvard University Press, 1992); Pnina G. Abir-Am, "The Politics of Macromolecules: Molecular Biologists, Biochemists and Rhetoric," *Osiris,* vol. 7, 1992, 164–91; and Kay, *The Molecular Vision of Life.*

8. By the early 1970s, progress in biology begin to suggest a shift from science to engineering. In 1971, before the phrase *genetic engineering* was coined, Graham Chedd reported on an immunology conference for *New Scientist and Science Journal,* May 13, 1971, 396. Chedd described how the immunologists thrashed out ways in which the immune system could be exploited—"how they might usher in the age of cellular engineering." Chedd went on to add a cautionary note, often lost in later, noisier times: "The conference's main problem was that even apparently firm theoretical ground turned out to be unexpectedly treacherous when it came to building bridges upon it."

9. For a general overview of the controversy over recombinant DNA, see Sheldon Krimsky, *Genetic Alchemy: The Social History of the Recombinant DNA Controversy* (Cambridge, Mass.: MIT Press, 1982).

10. The most evocative contemporary reportage on Asilomar is from M. Rogers, "The Pandora's Box Conference," *Rolling Stone,* June 19, 1975, 28, which is included in the documents compiled in James Watson and John Tooze, *The DNA Story* (San Francisco: W.H. Freeman, 1981).

11. Watson and Tooze, *The DNA Story,* ix.

12. Jeremy Rifkin and Ted Howard, "Who Shall Play God?" (New York: Dell, 1977). In an October 1986 interview in MIT's *Technology Review* (p. 38), David Baltimore commented on Rifkin and the controversy: "Many university scientists feel the whole Asilomar process burned them badly, and they don't want to see anything like that again. What started as an intellectual investigation ended up becoming a political circus. And the most outrageous moment. . . was generated by Jeremy Rifkin. . . . I understand that sort of thing as just the excesses of the process, and that the process is basically correct. But I can understand my colleagues who say, 'You're an idiot for opening up.'"

13. Stephen Hall, *Invisible Frontiers: The Race to Synthesize a Human Gene* (New York: Atlantic Monthly Press, 1987), 26.

14. For an overview of the patent controversy, see Charles Weiner, "Uni-

versities, Professors and Patents: A Continuing Controversy," *Technology Review*, February 1986, 32.

15. On the background to the Whitehead controversy, see "A $127 Million Gift Horse," *Newsweek*, October 12, 1981, 87; and Martin Kenney, *Biotechnology: The University-Industrial Complex* (New Haven, Conn.: Yale University Press, 1986). For a later perspective on a functioning Whitehead, see Natalie Angier, *Natural Obsessions: The Search for the Oncogene* (New York: Houghton Mifflin, 1988); and Corie Brown, "The Institute Where Some of Biotech's Best Brains Storm," *Business Week*, July 11, 1988, 98.

16. With many different, overlapping rules, disputes still regularly occur. See Christopher Anderson, "Hughes' Tough Stand on Industry Ties," *Science*, February 12, 1993, 884, particularly two sidebars, "Conflict Confusion: Five Views on Equity," p. 884, and "Federal Conflict Rules Nearing Completion," p. 885.

17. Eli Ginzberg and Anna Dutka, *The Financing of Biomedical Research* (Baltimore, Md.: Johns Hopkins University Press, 1989).

18. Sandra Panem, *The Interferon Crusade* (Washington, D.C.: Brookings Institution, 1984); and Robert Teitelman, *Gene Dreams: Wall Street, Academia and the Rise of Biotechnology* (New York: Basic Books, 1989), particularly "Raising Consciousness," 27–35.

19. George Gilder, *The Spirit of Enterprise* (New York: Simon & Schuster, 1984), 258.

20. Description of biomania in Teitelman, *Gene Dreams*, 11–14.

21. Genentech annual report, 1981, 2.

22. Howard E. Greene, Jr., "Hybritech Incorporated: Presentation to the Financial Community," *Hybritech*, November 1982, 1.

23. On the issue of biotech boutiques, see Robert Teitelman, "Biotech 'Boutiques' May Yield Greater Financial Rewards," *Oncology Times*, March 1989, 3; and Robert Teitelman, "Biotech Financing: Lessons from the Computer Industry," *Oncology Times*, April 1989, 5.

24. For a very rosy profile of Genentech prior to t-PA, see "Biotech's First Superstar," *Business Week*, April 14, 1986, 68. For generous views of the t-PA market, see Gene Bylinsky, "The New Assault on Heart Attacks," *Fortune*, March 31, 1986, 80; and "Biotech's Battle Over Heart Drugs," *Business Week*, December 15, 1986. There were warning signs of problems, such as those discussed in "Tissue Plasminogen Activator: Will It Fulfill Its Promise?" *New England Journal of Medicine*, October 17, 1985.

25. For a discussion of the intersection between regulation and commerce, see James Dickinson, "Storm Over t-PA: Wall Street Misses Fine Points of Approval," *Pharmaceutical Executive*, September 1987, 24.

26. Marilyn Chase, "As Genentech Awaits New Test of Old Drug, Its Pipeline Fills Up," *Wall Street Journal*, April 30, 1992, 1.

## Chapter 11: Gravity's Rainbow

1. Russell Mitchell, "Can Cray Reprogram Itself for Creativity?" *Business Week,* August 20, 1990, 86.
2. George Gilder, *Microcosm* (New York: Simon & Schuster, 1989).
3. Ludwig von Mises, *Bureaucracy* (New Haven, Conn.: Yale University Press, 1944); and Joseph Schumpeter, *Capitalism, Socialism and Democracy* (New York: Harper Torchbooks, 1975).
4. Peter Drucker, *The Practice of Management* (New York: Perennial Library, 1986).
5. Ralph Hower and Charles Orth III, *Managers and Scientists: Some Human Problems in Industrial Organizations* (Boston: Harvard University, 1963), 19.
6. John Kenneth Galbraith, *The New Industrial State* (Boston: Houghton Mifflin, 1967), 16.
7. Peter Drucker, *Innovation and Entrepreneurship* (New York: Harper & Row, 1985), 253.
8. Peter Drucker, *The Post-Capitalist Society* (New York: Harper Business, 1993). For a skeptical survey of management theorists, including Drucker, see Stephen P. Waring, *Taylorism Transformed: Scientific Management Theory since 1945* (Chapel Hill, N.C.: University of North Carolina Press, 1991). As this study points out, Drucker's most famous, contribution—management by objective—is not commonly employed today.
9. See Edward Denison, *Accounting for Slower Economic Growth* (Washington, D.C.: Brookings Institution, 1979). For a later survey of explanations and proposals on slowing productivity growth, see Michael L. Dertouzas, Richard K. Lester, Robert M. Solow, and the MIT Commission on Industrial Productivity, *Made in America: Regaining the Productive Edge* (Cambridge, Mass.: MIT Press, 1989).
10. Stephen Roach, "Services under Siege—the Restructuring Imperative," *Harvard Business Review,* September–October 1991, 86.
11. Stephen Roach, "White Collar Shock: Services Face the 1990s" (New York: Morgan Stanley, June 8, 1992).
12. Stephen Roach, "On the Road to Deliverance" (New York: Morgan Stanley, April 10, 1993).
13. These charges are summed up in two reports: Michael Porter, *Capital Choices: Changing the Way America Invests in Industry* (Washington, D.C.: Council on Competitiveness, 1993); and The Twentieth Century Fund Task Force on Corporate Governance, *Who's Minding the Store* (New York: Twentieth Century Fund, November 1992). For a general overview of the controversy, see Robert Teitelman, "Wall Street and the New Economic Correctness," *Institutional Investor,* February 1993, 36.
14. Paul Krugman, *The Age of Diminished Expectations: U.S. Economic Policies in the 1990s* (Cambridge, Mass.: MIT Press, 1989).

15. George Gilder, *The Spirit of Enterprise* (New York: Simon & Schuster, 1984), 44.
16. Gilder, *Microcosm*, 382–83.
17. See James R. Beniger, *The Control Revolution: Technological and Economic Origins of the Information Society* (Cambridge, Mass.: Harvard University Press, 1986), 4–5, table 1.1.
18. Ibid. Beniger traces his notion of a "control revolution"—information as the key measure of control—back to the early Industrial Revolution. Indeed, he argues that the increasing mastery, or control, of information is synonymous with industrialization. Beniger credits economist Fritz Machlup for coining the term *information society* and agrees that the information society entered a new phase with the advent of the microprocessor in the 1970s.
19. See David Freedman, "A Clouded Future for IBM Research," *Science*, April 23, 1993, 480–81.
20. See Elizabeth Corcoran, "Computing's Controversial Patron," *Science*, April 2, 1993, 20–23. For a brief overview of government failures in choosing technologies, see Eliot Marshall, "Technology Boosting: A Checkered History," *Science*, March 1993, 1817.
21. Michael Porter persuasively argues the case for the benefits of intense domestic competition in his *Competitive Advantage of Nations* (New York: Free Press, 1990). In his study on corporate governance, *Capital Choices*, Porter argues that mechanisms need to be developed to encourage investment not in smaller high-tech companies, which may well be overfunded, but among the great range of corporations that have entered middle age but are not yet mature.
22. Charles Ferguson and Charles Morris, *Computer Wars: How the West Can Win in a Post-IBM World* (New York: Times Books, 1993), 236–51, make many of these same points, arguing that the federal government must create an investment environment, particularly for basic research, and that it must discriminate between mainstream commodity computer products, such as memory chips, and what they call "architectural products," such as microprocessors and custom chips. They also make an excellent argument for a very limited form of managed trade: protect only those products for which there is a real threat of a cartel, as there was in Japanese memory chips in the late 1980s.
23. Political realities push and pull here. Votes come from small businesses. And as a self-described populist, Clinton is not about to challenge publicly the current orthodoxy that creativity and innovation come from small businesses. On the other hand, Clinton also won support from some twenty-eight Silicon Valley CEOs during the campaign—all from fairly large companies, a number of them nominal Republicans; and he tried to appoint John Young, the former CEO of Hewlett-Packard (and a Republican) to be his commerce secretary.

There is a realpolitik at work here, as there is in the emphasis on applied science. In a lagging economy, large companies have jobs today; small businesses only offer the promise of jobs tomorrow.

24. Jon Cohen, "A 'Manhattan Project' for AIDS?" *Science,* February 19, 1993, 112. See particularly the sidebar to that story, "History's Winners and Losers," 1114.

25. MITI itself recognizes the difference between following and leading. For the past decade or so the agency has tried to encourage broader technological initiatives ("strategic research" in the current parlance)—notably, with the Fifth-Generation computing project. Now, searching for new ways to justify its existence (much like DARPA), MITI has launched a broad, interdisciplinary basic research effort in a number of fields. See Fred Myers, "MITI Moves into Basic Research," *Science,* December 11, 1992, 1727.

# INDEX